County Report 52

# GEOLOGY AND MINERAL RESOURCES OF PIKE COUNTY, PENNSYLVANIA

by **W. D. Sevon**
Pennsylvania Geological Survey
**Thomas M. Berg**
Ohio Geological Survey
**Lane D. Schultz**
Allan A. Meyers, Inc.
**George H. Crowl**
Ohio Wesleyan University

PENNSYLVANIA GEOLOGICAL SURVEY

FOURTH SERIES

HARRISBURG

1989

ISBN: 0-8182-0120-7

ADDITIONAL COPIES
OF THIS PUBLICATION MAY BE PURCHASED FROM
STATE BOOK STORE, P. O. BOX 1365
HARRISBURG, PENNSYLVANIA 17105

# *Dedication*

## GEORGE H. CROWL

1910–1987

George H. Crowl contributed to the maps and report on Pike County in many ways. He mapped the glacial geology of the Delaware River valley. He assisted in mapping the remainder of the surficial deposits of the county. He reviewed all of the manuscript and helped strengthen it. But more than that, George contributed to the professional development of the other authors. He kept us honest. He made us think. The years we spent with George at the East Lawn Research Center were rich in comradery and intellectual growth. Thank you George. It was fun.

# PREFACE

The nature and occurrence of rocks and unconsolidated surficial deposits in Pike County, Pennsylvania, are described in this report. The rocks are a diverse assemblage of claystone, siltstone, sandstone, and conglomerate, whereas the unconsolidated deposits consist mainly of glacial till and various glacial- or stream-deposited sands and gravels.

Each of the materials mapped in the county has its own characteristics which affect the way the land and its resources may be utilized by man. In an area that is subjected to continuing demands for proper land utilization, water supply, waste disposal, and resource availability, the information provided herein should be of value to planners, homesite developers, engineers, water-well drillers, consulting geologists, and all who are interested in the proper management of the resources of the county.

To meet the needs mentioned above, this report contains data and interpretations of the physical characteristics, thickness, engineering characteristics, and mineral-resource potential of the materials mapped in the county. Although the report will serve as a guide to geologic problem solving, more detailed site investigations may be desirable where costly decisions must be made.

# CONTENTS

# ILLUSTRATIONS

## FIGURES

## PLATES

(in envelope)



## TABLES

# GEOLOGY AND MINERAL RESOURCES OF PIKE COUNTY, PENNSYLVANIA

by

W. D. Sevon, Thomas M. Berg, Lane D. Schultz, and George H. Crowl[1]

## ABSTRACT

Pike County, Pennsylvania, consists of about 545 square miles of diversified terrain located mainly in the Appalachian Plateaus physiographic province. The Delaware River valley and the Pocono Plateau escarpment are the two most prominent features of the topography. Rock units defined and mapped in the area are the lithified sediments that were deposited in a variety of offshore marine, marginal marine, and fluvial environments associated with the filling of the Appalachian basin by the prograding Catskill deltaic complex and coastal alluvial plain during the Devonian Period.

Surface-exposed bedrock units in Pike County constitute an approximately 9,500 foot vertical sequence that includes the black shale of the Marcellus Formation, the dark-gray siltstone and claystone of the Mahantango and Trimmers Rock Formations, and the gray sandstone and conglomerate and grayish-red and gray siltstones of the Towamensing, Delaware River, Lackawaxen, and Poplar Gap and Packerton Members of the Catskill Formation.

Surficial deposits occur throughout the county and include Olean Till, Olean ice-contact stratified sand and gravel, shale-chip rubble, boulder accumulations, alluvium and Olean outwash, alluvium and alluvial fans, swamp deposits, peat, artificial fill, and debris-avalanche chutes.

The basic geologic structure of Pike County is a homocline having northwest dips of a few degrees and a few small kink-band folds.

All of the rocks in the county have engineering properties that are generally favorable for construction purposes. The aquifer potential of most rock units is satisfactory. These rocks are all resistant to weathering and excavation, but numerous joints contribute to their natural breakup by frost wedging. Surficial deposits have less desirable engineering and aquifer properties, but are easy to excavate.

Currently exploited and potential mineral resources include sand and gravel, crushed rock, peat, bluestone (flagstone), and shale-chip rubble.

[1]Deceased.

## INTRODUCTION

### GENERAL

Pike County consists of about 545 $mi^2$ (square miles) of dominantly (91 percent) forested land in northeastern Pennsylvania. The county is bounded on the northeast by New York and on the southeast by New Jersey, and is separated from those states by the Delaware River. Part of the northwestern boundary is along Wallenpaupack Creek and in Wallenpaupack Lake. The remainder of the boundaries are surveyed. Nineteen percent of the area is State Forest land, 6 percent is State Game land, 1.6 percent is State Park land, and 5 percent is Federal Recreation area.

The county is crossed from east to west by Interstate Route 84, which gives easy access to New York City and the New England states, and from northeast to southwest by U.S. Route 209, which gives access to upstate New York and southeastern Pennsylvania. The remainder of the county is largely accessible by a network of paved and unpaved secondary roads, but some of the county is accessible only to nonvehicular traffic.

The population of Pike County in 1980 was 18,271, a figure that represented a 54.6 percent increase since 1970 and a 99.5 percent increase since 1960 (U.S. Department of Commerce, Bureau of Census, 1981). County population estimates for July 1, 1981 and 1982 (from Federal State Cooperative Program for Population Estimates), show that Pike County has the highest growth rate in the state, 7 percent. Accompanying this growth in population was a 74.8 percent increase in the number of housing units between 1970 and 1980. The largest boroughs in the county are Milford, the county seat, and Matamoras, both situated along the Delaware River in the extreme northeastern part of the county. Aside from a minor amount of farming, minimal light industry, and the usual businesses, most of Pike County is utilized for local residences and recreation.

The climate of Pike County consists of mild summers, cold winters, and an ample amount of precipitation. Daily, weekly, and seasonal variations and extremes are the norm. Average daily temperatures range from a minimum of 15 °F in January to a maximum of 78 °F in July. Precipitation is normally spread evenly throughout the year and averages about 51 inches per year. More detailed information on the climate is available in Taylor (1969) and Davis (1989). Normal precipitation and snowmelt conditions in Pike County cause minimal problems along streams, but, because the county lies within the inland hurricane track and is marginal to the coastal hurricane track, periodic floods result from excess precipitation produced by hurricanes. Information on flooding in Pike County was presented by Crowl (1971).

## PREVIOUS WORK

Some geologic information about Pike County was included in the summary report of the First Geological Survey of Pennsylvania (Rogers, 1858), but the first overall review of the geology of the county was that of White (1882). White identified the major rock-stratigraphic subdivisions within the area, discussed the glaciated character of the county, and mentioned specific details for many locations. His report, insightful as it is, does not provide adequate information for the questions of today.

Prosser (1894) reported on fossil collections made at various locations adjacent to the Delaware River in the Mahantango and Trimmers Rock Formations, but added no new knowledge about the geology of the county. Willard (1936) and Willard and others (1939) established a stratigraphic framework differing from that of White for Pike County and presented a generalized geologic map for the area. However, work done for the present report indicates that Willard's stratigraphy is not valid and that his map is based more on presumed correlations than on field mapping or real rock-stratigraphic contacts.

More recently, Fletcher and Woodrow (1970) mapped the geology of the northeastern part of the county around Matamoras, Milford, and Shohola. They established more refined rock-stratigraphic units and prepared a useful geologic map for the area. Crowl (1971) mapped the surficial deposits of the Delaware River valley between Matamoras and Bushkill and provided the first modern work on the glacial deposits in the county. The soils of the county were mapped by Taylor (1969), and general comments on glaciation of the area were made by Sevon and others (1975).

## PROCEDURES

Mapping in Pike County was done by standard field methods and extensive utilization of aerial photographs. Color aerial photographs at a scale of about 1:24,000 and taken in March 1973 were used for this project. The value of color aerial photography was discussed by Berg and Sevon (1974). Bedrock data were collected at all easily accessible localities, and walking traverses into less accessible areas were made wherever necessary. Bedrock contacts were located at specific points in the field, drawn with use of stereopairs of aerial photographs, and adjusted to topographic contours relative to local bedrock strike and dip. Contacts of surficial units were drawn on aerial photographs using known ground control and aerial-photograph interpretation. Surficial geologic maps produced at 1:24,000 scale were field checked, and revisions were made using the aerial photographs and the Pike County soils report (Taylor, 1969) prior to compilation at 1:50,000 scale.

Seventy-eight samples of till were collected throughout the county, and three samples were collected from neighboring counties. The proportions of gravel and sand-silt-clay were determined for each sample by weighing the bulk sample and the washed-and-dried fraction greater than 2 mm. The relative amounts of sand, silt, and clay were determined by the hydrometer method of Bouyoucos (1928).

Forty samples of sand and of sand and gravel were collected throughout the county. These samples were processed at the Pennsylvania Department of Transportation Testing Laboratory in Harrisburg. The tests performed were size analysis by sieving, specific gravity, absorption, and freeze-thaw breakup using a 5 percent alcohol-water solution. Pebble lithology identification and counts were done at the Pennsylvania Geological Survey.

Two hundred four thin sections of rocks collected from the Catskill Formation throughout the county were studied through a petrographic microscope. The average grain size of each sample was estimated visually from the thin section, and the mineral composition of each thin section was determined by component identification at 500 points.

Fracture traces interpreted from and drawn on color aerial photographs were projected onto 1:24,000-scale topographic maps. None of the interpreted fracture traces were checked in the field.

## WORK RESPONSIBILITY

The bedrock and surficial geology of the Pike County parts of the Culvers Gap, Flatbrookville, and Lake Maskenozha quadrangles was mapped by Schultz in 1972. Schultz mapped the Pecks Pond and Twelvemile Pond quadrangles in 1973. Berg and Sevon did most of the field work in the remainder of the county in 1973 and 1974. Intermittently over the course of several years, Sevon compiled the surficial geologic map by combining aerial-photograph interpretations and available field information. Crowl and Sevon spent two months in 1979 studying the surficial geology of the county and modifying the compiled surficial geologic map. Both Berg and Sevon spent short periods of time obtaining additional bedrock information in 1980, 1981, and 1982. Except for the sections on the Marcellus, Mahantango, and Trimmers Rock Formations, which were written by Berg, the text was written by Sevon. All authors have reviewed the text.

## ACKNOWLEDGEMENTS

The authors thank the following summer assistants who helped with the field work: Kimball Bodoia[1], Joseph Hebelka, Linda Main, and Timothy Troutman. They further thank Linda Main, who did most of the sand and gravel analyses and many of the till analyses. Glascow Rector of Geological Sample Log Company kindly gave permission to publish information derived from a lithologic log (Geolog) of a Texaco deep well. The authors also thank the many residents of Pike County who allowed them access to private property.

# GEOMORPHOLOGY

Most of Pike County lies in the Glaciated Low Plateau section of the Appalachian Plateaus province and is characterized by a relatively uniform topography of low to moderate relief. The western corner of the county is in the Glaciated Pocono Plateau section of the province and is similar in character except for a steep escarpment which marks its southern boundary. This escarpment loses definition north of High Knob, and the boundary between the Low Plateau and the Pocono Plateau becomes somewhat arbitrary. A narrow area of moderately to steeply dipping rocks paralleling the Delaware River on the south-

[1]Deceased.

eastern side of the county is included in the Appalachian Mountain section of the Ridge and Valley province.

The terrain of Pike County has been categorized by the authors on the basis of combined geologic, topographic, and drainage characteristics. Figure 1 shows terrain areas, and their general characteristics are given below.

AREA 1: Undulating upland topography developed mainly on resistant, gray, flaggy sandstone and local areas of glacial till. Sand and gravel occur along streams and some valley sides. Drainage is poorly developed; the area has numerous peat bogs. Moderate to low slopes generally have less than 100 feet of relief in upland areas; steep slopes generally have over 400 feet

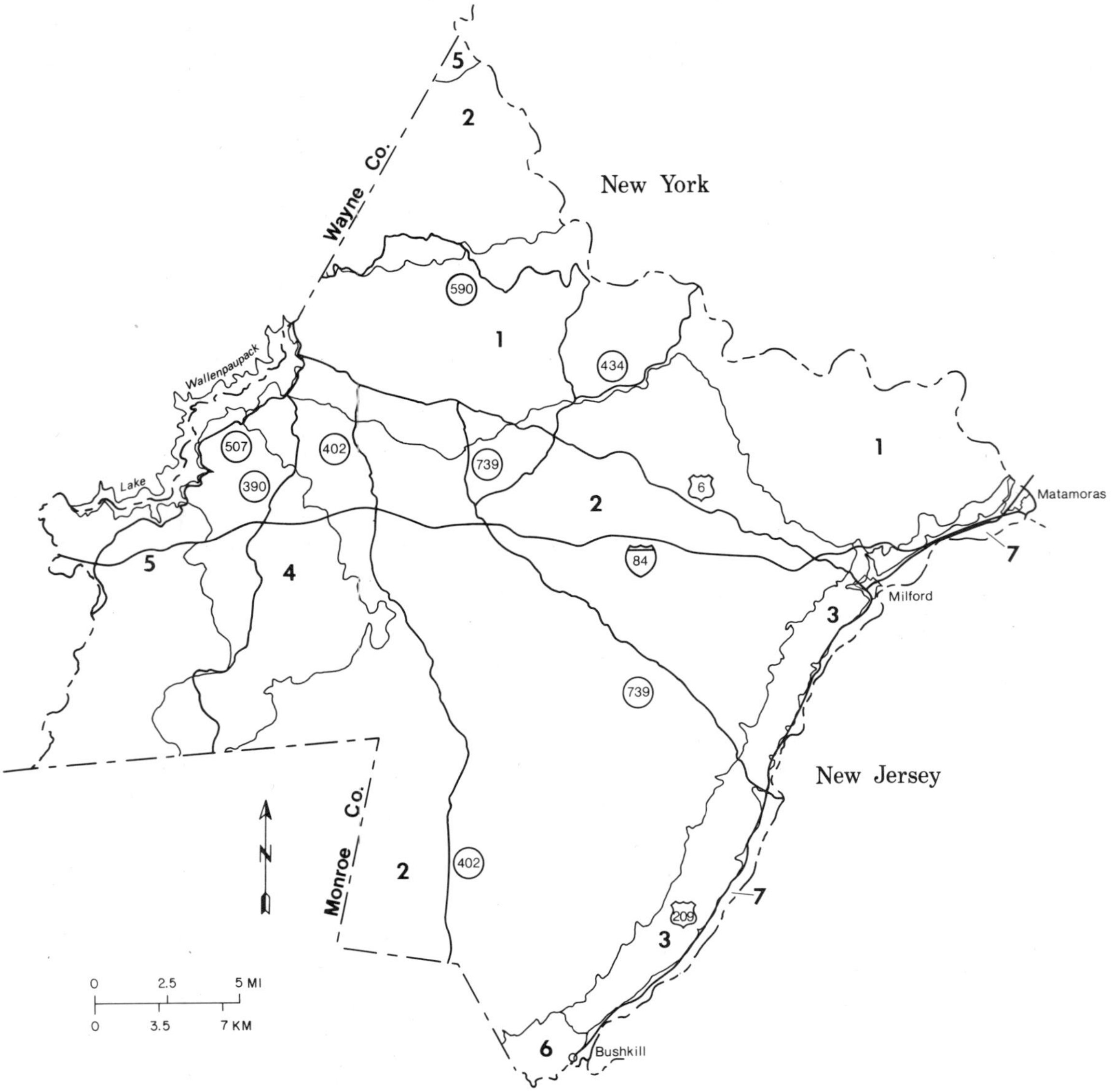


Figure 1. Pike County terrain subdivisions. An explanation for each area is given in the text.

of relief along major streams. Soils are deep, well drained, and loamy to very stony.

AREA 2: Undulating topography composed of numerous ledges of gray sandstone alternating with low areas of red siltstone and shale. Extensive areas of glacial till and local mounds of sand and gravel occur. Drainage is very poorly developed; the area has many lakes, peat bogs, and swamps. Moderate to low slopes generally have less than 100 feet of relief. Soils are deep, moderately well drained, and loamy to stony.

AREA 3: Linear topography developed on moderately steep, northwest-dipping, dark-gray siltstone and shale. Minor, local accumulations of glacial till occur. A steep slope having 400 to 500 feet of relief faces the Delaware River valley and is frequently masked by shale-chip rubble. Numerous rock ledges and steep rock slopes alternate with narrow, linear valleys generally having less than 100 feet of relief. The area is moderately well drained by steep-gradient tributaries to the Delaware River. Moderately deep, well-drained soils are mixed with small, local areas of very shallow or no soil.

AREA 4: Upland area developed on resistant gray sandstone and conglomerate and some less resistant red siltstone and shale. Low sandstone ledges commonly alternate with topographically lower areas of glacial till and boulder accumulations. Drainage is poorly developed. The southern edge of the area is marked by the steep Pocono Plateau escarpment, which has a relief up to 600 feet in the southwestern sector, decreasing to less than 50 feet in the northeastern sector. The upland plateau surface generally has low slopes and local relief less than 50 feet. Soils are mainly deep, well drained, and stony.

AREA 5: Rolling topography developed on glacial till, gray sandstone, and red siltstone and shale. The area includes extensive deposits of glacial till, local accumulations of sand and gravel, and minor occurrences of rock outcrops. A well-integrated dendritic drainage system occurs with local undrained or poorly drained depressions and numerous small lakes. Slopes are moderate, and local relief is generally between 100 and 200 feet. Soils are well drained to poorly drained and channery.

AREA 6: Dissected, slightly linear topography developed on dark-gray siltstone, black shale, glacial till, and sand and gravel. The till and sand and gravel deposits are scattered and variable in thickness. A well-developed rectangular drainage system is modified somewhat by glacial deposits. Moderate to steep slopes have local relief generally around 100 feet. Soils are shallow to deep and poorly to well drained.

AREA 7: Flat or moderately sloping topography developed on glacial or alluvial gravel, sand, and clay deposits. These valley deposits are associated with major drainage. Slopes are mainly low and locally moderate. Local relief is generally less than 10 feet, but in some places is up to 50 feet. Soils are generally deep and well drained.

## SUBSURFACE STRATIGRAPHY

The following brief descriptions of rocks occurring in the subsurface of Pike County are derived from the deep well, Pennsylvania Department of Forests and Waters #1, drilled by Texaco, Inc., in 1971, in Blooming Grove Township at 41°25′02″N/75°08′29″W (Hawley quadrangle). This well is the deepest stratigraphic test within Pike County. Lithologic descriptions are generalized from Geolog, and the formation boundaries are interpreted from the lithologic descriptions and geophysical logs. The names of all units below the Oriskany are from the Geolog and may or may not be correctly applied.

### ORDOVICIAN

BEEKMANTOWN GROUP: 120 feet penetrated. The interval consisted of light- to medium-dark-gray, dense, lithographic to microcrystalline dolomite and some beds of very dolomitic, dark-gray siltstone and dolomitic, brownish-gray, coarse- to very coarse grained sandstone having rounded sand grains. The dolomite was partly very siliceous, slightly silty, partly calcareous, slightly pyritic, and locally fossiliferous (ostracodes).

BLACK RIVER FORMATION: 40 feet thick. The formation was mainly dark-brownish-gray to dark-gray, lithographic to rarely microcrystalline, dolomitic, pyritic limestone. There was also a trace of lithographic dolomite, and soft, medium-dark-gray bentonite occurred at the base.

TRENTON GROUP: 152 feet thick. The interval was made up of light- to dark-gray, fragmental, pyritic, fossiliferous, slightly argillaceous limestone interbedded with very calcareous, locally fossiliferous, black shale. Brachiopods, bryozoans, crinoid columnals, and ostracodes were noted.

UTICA FORMATION: 568 feet thick. The interval consisted of very dark gray to black, platy to massive, in part slightly calcareous, in part slightly pyritic shale containing stringers of dark-brownish-gray to black, lithographic to cryptocrystalline limestone.

MARTINSBURG FORMATION: 280 feet thick. The formation comprised very dark gray to black, fairly hard, platy, slightly silty shale interbedded with dark-gray, slightly argillaceous, micaceous, dolomitic, pyritic siltstone. The siltstone graded into very fine grained, dark-gray to black sandstone in some places. Shale in the lower half was slightly calcareous in part and had traces of dark-gray to black, lithographic limestone.

## SILURIAN

TUSCARORA (SHAWANGUNK) FORMATION: 767 feet thick. The formation consisted of light-gray to pale-grayish-green, very fine to coarse-grained, moderately sorted, siliceous to very slightly dolomitic sandstone having angular to subrounded grains and containing streaks of pale-grayish-green to black shale. The interval included scattered quartz pebbles and was locally conglomeratic in all but the upper 250 feet.

ROSE HILL FORMATION: 441 feet thick. The interval contained pale-green to very light to moderate-greenish-gray, fairly soft, platy shale having a trace of black, lithographic limestone; pale-green, very fine to fine-grained sandstone; some massive, hard, black shale in the middle; and some stringers of dark-brown to black, pyritic, lithographic limestone in the lower part.

KEEFER FORMATION: 247 feet thick. The lower part of the interval comprised grayish-red to pale-green to light-gray shale and some brownish-gray, lithographic limestone. The middle part included light-gray to light-greenish-gray, very fine to coarse-grained, siliceous to slightly feldspathic, in part very slightly hematitic sandstone having angular to rounded grains and containing some streaks of medium-dark-gray to black and red shale. The upper part consisted of grayish-green to greenish-gray, very fine to fine-grained, fairly well sorted, very slightly hematitic, very slightly pyritic, slightly dolomitic, slightly feldspathic sandstone having angular to rounded grains, containing scattered quartz pebbles, and interbedded with medium-gray, soft, platy shale; and a trace of gray to black, lithographic limestone.

CLINTON FORMATION: 155 feet thick. The formation was dominantly light- to medium-dark-gray to greenish-gray, very fine to fine-grained, fairly well sorted, silty, slightly feldspathic, slightly pyritic sandstone having angular to subangular grains and interbedded with pale-grayish-green to greenish-gray, platy, soft silty shale. A trace of light-gray to black, lithographic limestone and some dark-gray to black, platy to massive, fairly hard, slightly micaceous, silty shale were found. There was also a trace of ostracodes and bryozoans near the base.

BLOOMSBURG FORMATION: 430 feet thick. The interval consisted of grayish-red, slightly argillaceous, slightly sandy, slightly micaceous siltstone; grayish-red to moderate-red shale containing stringers of red, fine-grained sandstone; light-gray to red, very fine to coarse-grained, poorly sorted, siliceous to feldspathic, slightly chloritic sandstone having angular to subrounded grains and some scattered, very coarse grains, and containing stringers of black shale; and traces of medium- to dark-gray, lithographic limestone.

WILLS CREEK FORMATION: 670 feet thick. The interval comprised light-grayish-green to grayish-red, platy to massive, slightly dolomitic shale having some red and gray mottling; some light-gray to light-greenish-gray, cryptocrystalline, very argillaceous, slightly sandy dolomite; and, at and near the base, black lithographic limestone and calcareous shale.

TONOLOWAY FORMATION: 721 feet thick. The formation included light-gray to white and some light-greenish-gray, cryptocrystalline to microcrystalline, partly very argillaceous to shaly, partly anhydritic, partly pyritic dolomite interbedded with light- to dark-gray, lithographic to cryptocrystalline limestone containing stringers of black shale.

KEYSER FORMATION: 263 feet thick. The interval contained the following lithologies from bottom to top: 25 feet of light- to dark-gray to grayish-brown, lithographic to cryptocrystalline, slightly argillaceous, slightly pyritic, in part slightly sandy limestone; 118 feet of dark-gray, calcareous, slightly fossiliferous shale interbedded with dark-gray, cryptocrystalline, very argillaceous limestone; 60 feet of dark-gray to black, lithographic to cryptocrystalline, in part slightly argillaceous limestone; and 60 feet of light-gray, lithographic to cryptocrystalline, shaly to very argillaceous dolomite interbedded with light-grayish-brown, lithographic, pyritic limestone.

## DEVONIAN

NEW SCOTLAND FORMATION: 65 feet thick. The formation consisted of dark-gray, chalky to fragmental to cryptocrystalline, in part slightly argillaceous, fossiliferous (scattered brachiopods and crinoid columnals) limestone and a trace of dark-gray translucent chert.

HELDERBERG GROUP: 420 feet thick. The interval contained dark-gray to black, cryptocrystalline, dense, slightly to moderately argillaceous, fossiliferous (brachiopods, gastropods, crinoid columnals, and ostracodes) limestone. In the upper 80 feet the limestone was interbedded with dark-gray to black, platy, hard, slightly pyritic, slightly micaceous, variably calcareous shale containing some brachiopods and ostracodes.

ORISKANY GROUP: 60 feet thick. The group contained dark-gray, platy, fairly hard, argillaceous, slightly calcareous, slightly pyritic, micaceous siltstone and a trace of dark-gray to black chert.

ESOPUS FORMATION: 310 feet thick. The interval consisted mostly of dark-gray, hard, slightly pyritic, slightly micaceous, noncalcareous shale interbedded with black, lithographic to cryptocrystalline, dense, slightly to very argillaceous, slightly silty limestone. The upper 50 feet contained dark-gray, hard, platy, calcareous (top) to noncalcareous (bottom), slightly fossiliferous (brachiopods, gastropods, and ostracodes) shale.

SCHOHARIE FORMATION: 280 feet thick. The interval comprised dark-gray, partly calcareous, fossiliferous shale interbedded with dark-gray to black, cryptocrystalline, dense, very argillaceous to shaly, fossiliferous limestone. Brachiopods, gastropods, *Tentaculites*, and ostracodes were noted.

BUTTERMILK FALLS LIMESTONE (ONONDAGA FORMATION): 320 feet thick. The limestone was dark gray to black, cryptocrystalline, dense, slightly to very argillaceous, fossiliferous, and in part slightly silty, and contained some light-gray chert. It was interbedded with dark-gray to black, platy, hard, slightly pyritic, calcareous shale.

The remainder of the rock units occurring in the Forests and Waters #1 well and in two other wells in Pike County are exposed at the surface within the county and are described in the section "Surface Bedrock Stratigraphy." Figure 2 presents a graphic summary of the rock units penetrated by the three exploration wells drilled in Pike County.

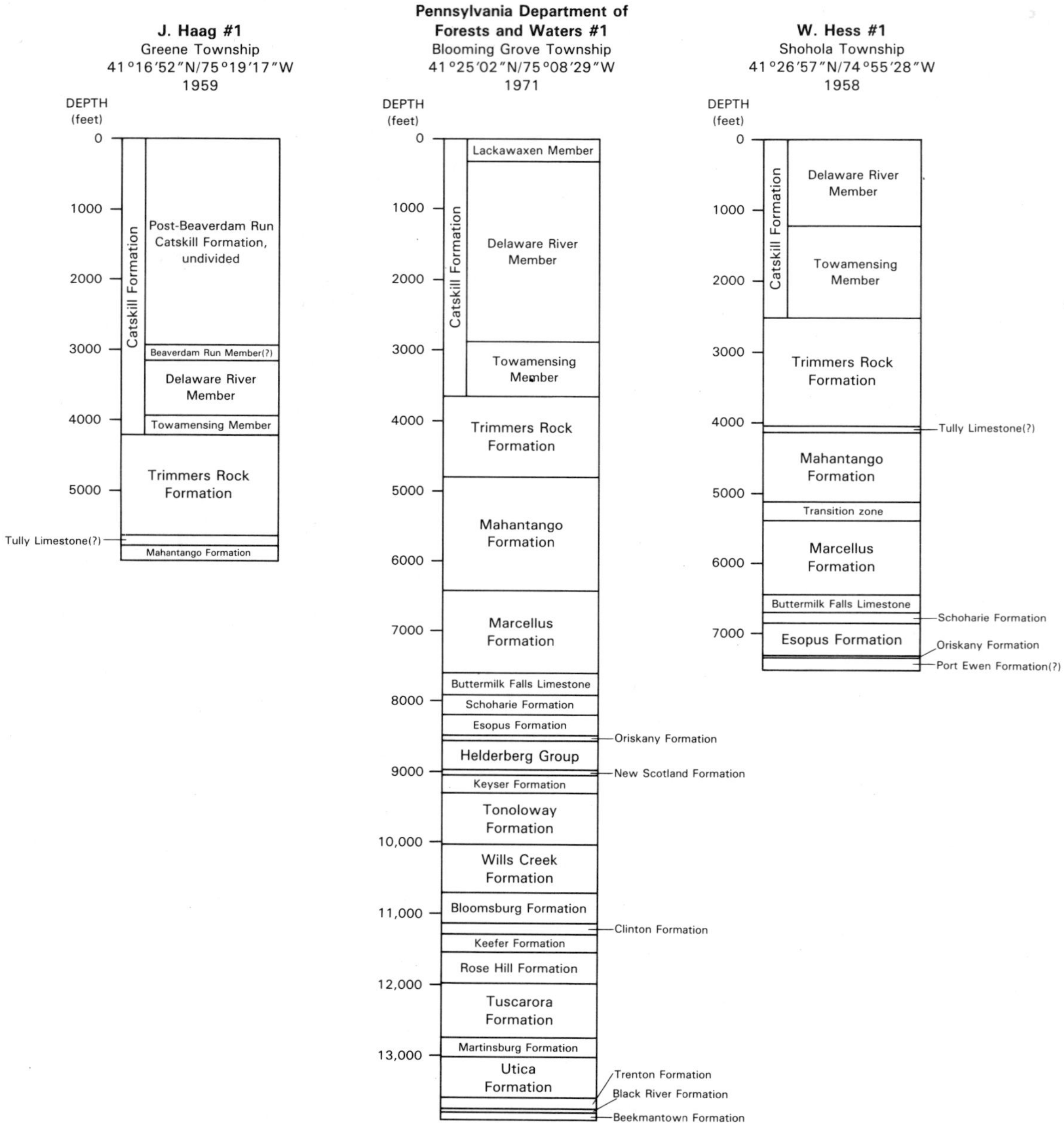


Figure 2. Stratigraphic columns of three gas/oil exploration wells drilled in Pike County. The rock-unit boundaries were interpreted by Sevon from sample descriptions and geophysical data.

# SURFACE BEDROCK STRATIGRAPHY

## INTRODUCTION

Rocks exposed at the surface in Pike County include, in ascending order, the Devonian Marcellus, Mahantango, Trimmers Rock, and Catskill Formations (Plate 1). The Trimmers Rock is subdivided (ascending) into the Sloat Brook and Millrift Members. The Catskill is subdivided (ascending) into the Towamensing, Delaware River, Lackawaxen, and Poplar Gap and Packerton Members, undivided. These rocks display an increase in grain size from clay in the Marcellus to pebbles in the Lackawaxen and Poplar Gap and Packerton. The interpreted environments of deposition for these rocks indicate a progressive upward change from distal marine basin (Marcellus) to alluvial plain (Lackawaxen and Poplar Gap and Packerton). Shoreline environments occur in the uppermost Trimmers Rock or lowermost Towamensing.

The changes in environments of deposition and sediment size reflect the northwestward advance of the Catskill deltaic complex and coastal alluvial plain that developed in the Middle and Late Devonian in response to uplift and erosion of the Acadian mountains, which then existed approximately 30 to 60 miles to the east of Pike County. As the mountains were uplifted, they were eroded, and many streams carried sediment to the marine sea that extended the length of the Appalachian basin. Coalescing deltas fed by these streams moved the junction of land and sea gradually northwestward until by the end of the Devonian the shoreline was in western Pennsylvania, and Pike County was the site of an alluvial plain traversed by braided and meandering streams. More information about the Catskill delta and its deposits is given in Sevon (1981), Thompson and Sevon (1982), and Woodrow and Sevon (1985).

Each of the surface rock units in Pike County is discussed in detail in the sections that follow. The stratigraphic nomenclature of some of the units has a complex history and is reviewed at the start of the discussion for those particular rock units. Figure 3 shows some of the stratigraphic nomenclature that has been used in Pike County and surrounding areas.

## MARCELLUS FORMATION

INTRODUCTION: The Marcellus Formation was mapped in the Bushkill quadrangle (which includes the southern corner of Pike County) by Alvord and Drake (1971), and their work is used herein. They divided the Marcellus into the Brodhead Creek Member (upper portion) and the Stony Hollow-Union Springs Member, undifferentiated (lower portion), following the usage of Willard (1938) and Cooper (1941). Only the dark-gray shale of the Brodhead Creek Member was mapped in Pike County by Alvord and Drake; the Stony Hollow-Union Springs Member is covered by alluvium and glacial deposits. Subdivision of the Marcellus was not attempted for this report because of lack of exposure.

FIELD OCCURRENCE: The Marcellus occurs at the surface as a relatively narrow belt across small parts of the Bushkill and Flatbrookville 7.5-minute quadrangles (see Plate 1). Glacial deposits and alluvium cover a large part of the outcrop belt. Because outcrops of the subjacent Buttermilk Falls Limestone occur close to the eastern edge of the Delaware River valley in New Jersey, the authors believe that the Marcellus is buried beneath glacial and alluvial deposits across most of the valley. The presumed boundary between the Marcellus and the overlying Mahantango Formation is shown as a concealed contact on Plate 1. No specific landform can be attributed to the Marcellus Formation in Pike County other than the steep slopes and bluffs above Bushkill, which are a result of sculpturing by glacial ice.

DESCRIPTION: The Marcellus Formation comprises slightly siliceous clay shale and silty clay shale, which are dark gray (N3)[1] to grayish black

[1]Color designations for bedrock units are from Goddard and others (1970).

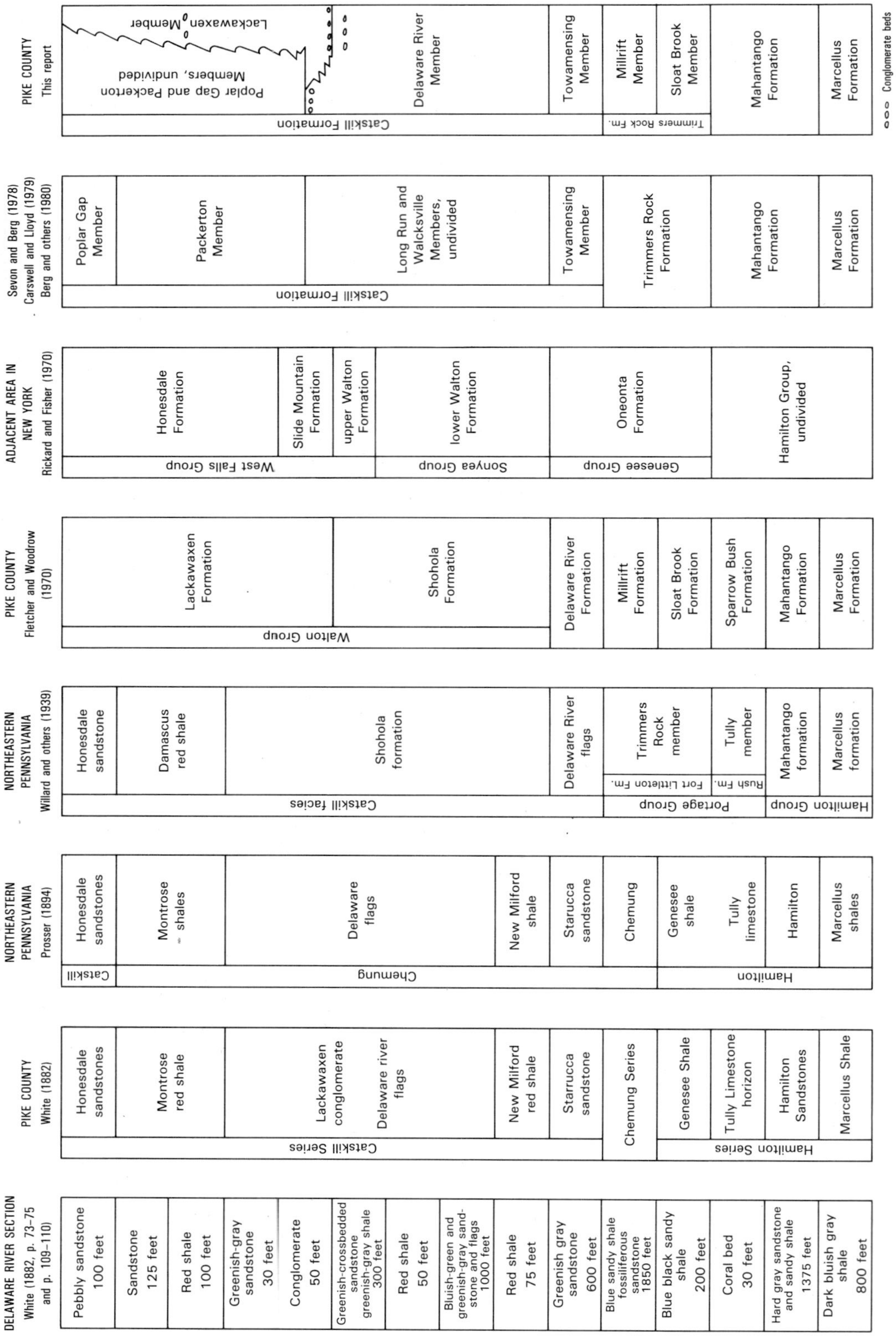


Figure 3. Stratigraphic nomenclature for Pike County and adjacent areas.

(N2). Some thin beds of argillaceous siltstone are present and are probably more common near the upper contact of the formation. Some beds are slightly micaceous, and there are traces of pyrite. Zones of nodules occur intermittently; they are generally arranged parallel to bedding and range in size from 2 cm to about 15 cm. The nodules vary in shape from subspherical to flat ellipsoidal. The nodules are composed of microcrystalline quartz and chlorite and contain minor amounts of muscovite (as determined by a diffractometer scan). Very small traces of feldspar (albite?), dolomite, calcite(?), pyrite, and siderite also occur in the nodules. In the Forests and Waters #1 well, stringers of black, lithographic, argillaceous limestone were reported in the lower half of the formation. Although not abundantly fossiliferous, the Marcellus does contain some marine fossils. Crinoid columnals and brachiopods are not uncommon; few trilobites have been noted.

Bedding in the Marcellus is obscured by cleavage, and the shale breaks up into small chips, plates, and flaky to splintery fragments. The formation is in most places reduced by weathering to deposits of shale-chip rubble. Systematic joints are not well developed in the Marcellus shale; joint surfaces tend to be irregular, discontinuous, and curved.

THICKNESS: Alvord and Drake (1971) reported that the Marcellus Formation has a maximum thickness of 950 feet. The Brodhead Creek Member constitutes the upper 750 feet of the formation. The 1,160-foot thickness reported in the Forests and Waters #1 well may be due to northwest thickening of the Marcellus, or it may be due to a slightly different interpretation of the Mahantango-Marcellus gradational contact.

CONTACTS: The lower contact of the Marcellus is not exposed in Pike County, but that contact with the underlying Buttermilk Falls Limestone is probably abrupt. The upper contact is gradational; it is placed at the change from the dark-gray, silty clay shale of the Marcellus to the lighter colored silt shale and siltstone of the Mahantango.

EXPOSURES: The Marcellus Formation has very limited exposure in Pike County because of its small outcrop area in the southern corner and because of cover by glacial deposits and alluvium. The best exposure is in a small shale quarry located about 1.2 miles west-southwest of Bushkill at 41°05′15″N/75°02′20″W.

ORIGIN: The fine sediments of the Marcellus were deposited in a marine environment having predominantly anoxic (negative Eh) bottom conditions. A lack of current features indicates that the muds were not affected by any well-developed circulation patterns and were deposited far below effective wave base. The dark-gray to grayish-black color of Marcellus shale is due to organic carbon and to the low oxidation state of iron in sulfides such as pyrite and marcasite (Schultz, 1973, p. 15). Pyrite precipitation under anoxic conditions is favored by pH values between 7 and 8.

## MAHANTANGO FORMATION

INTRODUCTION: The medium-dark-gray siltstone and silty shale of the Mahantango Formation were mapped along the Delaware River valley from Matamoras to the vicinity of Dingmans Ferry by Fletcher and Woodrow (1970). They subdivided the Mahantango into three informal members, but mapped their "upper member" for only 5 miles southwest of Milford. Fletcher and Woodrow (p. 8) stated that their "middle member" included gray siltstone, fine-grained sandstone, and shale. The present authors could not subdivide the Mahantango Formation as Fletcher and Woodrow did. Although there is a subtle upward increase in grain size and bedding thickness, no clear-cut internal members could be defined for this county report; only coarse-grained siltstone, rather than sandstone, was observed. Alvord and Drake (1971) subdivided the Mahantango into two informal members at the position of a fossiliferous siltstone, which they called the "Centerfield Reef." This bed does not extend far into Pike County and could not be used to subdivide the formation.

FIELD OCCURRENCE: The Mahantango Formation occurs in a long, approximately 1.5- to 2-mile-wide belt parallel to the course of the Delaware River, along the southeastern side of Pike County (see Plate 1). The shale in the lower part of the formation is well exposed along the Delaware Valley, where it has been sculpted to form steep cliffs and bluffs by the river and glacial ice. The terrain north and west of the valley is characterized by many small, subparallel ridges or ledges, reflecting the monoclinal, moderately dipping siltstone beds of the upper part of the Mahantango. The resulting drainage is very close to a trellis pattern, but is modified somewhat by the presence of thin glacial-till deposits and thick stratified-drift deposits.

DESCRIPTION: The Mahantango Formation comprises a long succession of interbedded claystone, clay shale, silt shale, and fine- to coarse-grained siltstone (Figure 4). The formation is dominantly siltstone (Figure 5); no sandstone was observed. There is a gross coarsening-upward aspect within the formation. Little lateral textural variation is discernible in Pike County. The shale and siltstone are predominantly dark gray (N3), but range to grayish black (N2), medium dark gray (N4), and rarely medium gray (N5). Weathering colors include grayish orange (10YR6/4), moderate yellowish brown (10YR5/4), and olive gray (5Y3/2) to light olive gray (5Y5/2). Mahantango shale is generally fissile to subfissile; fissility (especially in the lower part of the formation) is commonly more the result of cleavage than of bedding. Mahantango siltstone is subfissile to nonfissile.

Typically, the shale and siltstone in this formation are parallel laminated and parallel bedded,

Figure 4. Interbedded claystone, shale, and thin siltstone in the lower part of the Mahantango Formation in an exposure in a bluff north of the intersection of Interstate Route 84 and U.S. Route 209 at 41°21′27″N/74°43′16″W. Note the siliceous nodules and bands (see arrows) at the level of the scale, which is divided in inches and centimeters.

Figure 5. Fresh exposure of the Mahantango Formation along Interstate Route 84 at 41°20′34″N/74°46′01″W, approximately 2 miles northeast of Milford. Note the massive character of the interbedded shale and siltstone, and the irregular joint surfaces. The roadcut is approximately 60 feet high.

with the shale thickly laminated to thin bedded and the siltstone medium to thick bedded. Bedding surfaces are in many places vague, giving many exposures a massive appearance (Figure 5). Thin silt laminae or zones of nodules are commonly the only indicators of bedding. Ripple crossbedding is present in some siltstone beds. The ripples are subtly asymmetric in cross section and sinuous in plan view. Rarely, the siltstone displays very small scale, low-angle crossbedding. Within the Mahantango, especially in the upper part, a few graded beds 15 to 20 cm thick occur (Figure 6), but these are less well defined than the graded beds of the overlying Trimmers Rock Formation. Rare small-scale load-cast structures also occur in the upper part of the Mahantango Formation.

The zones of nodules that occur within the Mahantango probably have a composition similar to the nodules in the Marcellus Formation. A microcrystalline quartz composition is likely; they do not react with dilute hydrochloric acid.

Figure 6. Graded bedding in the Mahantango Formation (closer view of the roadcut shown in Figure 5). The scale is divided in 10-cm intervals. The arrow is at the base of the graded unit.

The nodules are generally flattened ellipsoids less than 4 cm thick (Figure 7).

Marine invertebrate fossils are common at various horizons in the Mahantango. These horizons are not mappable units, as they are in Monroe and Carbon Counties to the southwest. Most notably, the very fossiliferous, persistent marine zone called the "Centerfield Reef" is not recognized in Pike County. Prosser (1894) listed numerous marine fossils that had been collected from exposures near Bushkill, Dingmans Ferry, and Milford in Pike County, as well as from exposures near Port Jervis, New York. In their report on the Pennsylvania portion of the Milford and Port Jervis 15-minute quadrangles, Fletcher and Woodrow (1970, p. 7–10) listed numerous marine invertebrate fossils occurring in the Mahantango, particularly in their "middle member." In the Lake Maskenozha quadrangle, Schultz (1973, p. 17) identified the fossil brachiopods *Mucrospirifer mucronatus* (Conrad), *Paraspirifer* sp., *Chonetes* sp., and *Rhipidomella penelope* (Hall) as being common at various horizons in the Mahantango Formation. Schultz reported the occurrence of the bryozoan *Fenestella* sp. He further listed *Grammysia* sp., *Nuculities oblongatus* (Conrad), and *Actinopteria decussata* (Hall) as the common bivalves. Schultz also in-

Figure 7. Fine-grained siltstone of the Mahantango Formation bearing flat, ellipsoidal siliceous nodules. The irregular jointing is typical of the lower part of the Mahantango. The exposure is along a road ascending a bluff north of the intersection of Interstate Route 84 and U.S. Route 209 at 41°21′25″N/74°43′25″W. The scale is divided in inches and centimeters.

dicated that cephalopods include *Michelinoceras* sp., and that gastropods include *Bembexia* sp. Fossil trilobites occur locally, and *Phacops* is the most common genus (Schultz, 1973, p. 17). *Greenops (Greenops) boothi* (Green) was found at 41°06′58″N/74°58′51″W in the Flatbrookville quadrangle by T. M. Berg.

The following list summarizes the authors' general field observations of fossils in the Mahantango Formation:

| *Fossil Group* | *Abundance* |
|---|---|
| Corals: small, rugose | Few to rare |
| Bryozoans | Few |
| Brachiopods: many genera | Very common |
| Gastropods: small, high-spired | Rare |
| Gastropods: large, low-spired | Few |
| Cephalopods: nautiloids, straight | Very few |
| Bivalves: wide variety | Common |
| Trilobites | Rare |
| Crinoid columnals | Abundant |
| Plant fragments: small | Few |
| Trace fossils: small burrows(?) | Very few |

The fossils as observed in the field were generally concentrated on specific bedding surfaces rather than homogeneously dispersed through a large vertical interval. In general, fossil communities appeared moderately diverse, although some appeared to be less diverse, particularly where larger gastropods were found. Rarely, as at the cliffs west of Matamoras, complete crinoid calyxes were found.

Mahantango shale disintegrates to chippy, platy, hackly, and splintery fragments. The splintery fragments are a result of the intersection of closely spaced bedding planes and closely spaced cleavage planes. Sevon and Berg (1979, p. 2–7) discussed the shale-chip-rubble deposits that are derived from very shaly formations in Pennsylvania. These accumulations of naturally fragmented Mahantango shale are discussed in detail in this report in the section "Shale-Chip Rubble" (p. 75–79).

MICROSCOPIC PETROLOGY: Mahantango siltstone contains a framework of quartz silt grains set in a composite binder of illite, chlorite, and microcrystalline silica. Much of the binder is probably authigenic. Limonite occurs as poorly defined patches in some places, and as better defined zones adjacent to fossil shells. The extensive development of the illite-chlorite-silica composite binder suggests a very low initial porosity. Any effective porosity is likely the result of fracturing.

THICKNESS: The Mahantango ranges from 2,450 feet in thickness in the eastern part of Pike County, the Milford-to-Matamoras area, to approximately 1,800 feet thick in the southern part of the county (Alvord and Drake, 1971), to 1,640 feet thick in the Pennsylvania Department of Forests and Waters #1 well in Blooming Grove Township, to 1,345 feet thick in the W. Hess #1 well near Shohola.

CONTACTS: The lower contact of the Mahantango Formation with the underlying Marcellus Formation is gradational over about 200 feet and is placed at the change from dark-gray, silty clay shale of the Marcellus to lighter colored silt shale and siltstone of the Mahantango. The upper contact of the Mahantango Formation with the overlying Trimmers Rock Formation (Sloat Brook Member) is placed at the first significant appearance of well-developed graded bedding (Bouma sequence), along with a preponderance of siltstone. The upper contact can also be inferred by the presence of load casts, which are typical of the Trimmers Rock. The first appearance of very fine grained sandstone is an alternative criterion for defining the upper contact, but sandstone beds are not ubiquitous in Pike County at this horizon. In this regard, the upper contact is gradational. Fletcher and Woodrow (1970, p. 10 ff.), however, recognized a 40-foot-thick unit overlying the Mahantango, which they named the Sparrow Bush Formation. Their Sparrow Bush comprises coquinite, fine-grained sandstone, and siltstone. The Sparrow Bush was thought to be equivalent in part to the Tully Limestone. The present authors could not trace a "Sparrow Bush" unit along strike from Fletcher and Woodrow's type area. Schultz (1973) did not recognize any mappable unit equivalent to the Sparrow Bush. Alvord and Drake (1971) did not recognize a coarse marker unit bearing coquinites at the Mahantango-Trimmers Rock contact and, in point of fact, described the contact as a gradation. The present authors include the apparently areally limited Sparrow Bush strata in the Mahantango, treating them as equivalent to the Tully Limestone.

EXPOSURES: There are profuse exposures of the Mahantango Formation in the appproximately 1.5- to 2.0-mile-wide belt parallel to and west of the Delaware River. Many ledges of northwest-dipping coarse siltstone protrude from beneath thin glacial till or colluvial deposits. The most spectacular outcrops may be seen at many locations along U.S. Route 209, which runs the length of the Delaware River valley from Bushkill to Matamoras. In part, these outcrops are man-made, occurring as large pits and roadside quarries where Mahantango shale-chip rubble has been excavated. Higher on the valley side are large natural bluffs and cliffs of Mahantango, which greatly add to the scenic beauty of the Delaware Valley in Pike County.

The best exposure of fresh, unweathered Mahantango shale and siltstone is in the large roadcut along Interstate Route 84, 2 miles northeast of Milford (see Figure 5); the east end of the roadcut is at 41°20′36″N/74°45′55″W, and the west end of the cut is at 41°20′32″N/74°46′11″W. A long series of small exposures providing a very good representation of lower Mahantango characteristics is located along a curving road ascending the southeast end of the bluff just north of the intersection of Interstate Route 84 and U.S. Route 209, west of Matamoras (see Figures 4 and 7). The southern end of the series of exposures is at 41°21′27″N/74°43′16″W. An excellent man-made exposure of Mahantango Formation siltstone and silt shale is in a large borrow area (Figure 8) on the west side of U.S. Route 209, 3.6 miles southwest of Milford, at 41°16′42″N/74°50′26″W. Hard, siliceous nodules and some marine fossils occur at this locality. Irregular, subvertical jointing in the Mahantango is particularly well displayed at this exposure.

ORIGIN: The Mahantango Formation in Pike County was deposited under marine conditions, in open water below normally effective wave

Figure 8. Exposure of Mahantango siltstone and silt shale in a borrow area along U.S. Route 209 southwest of Milford, at 41°16′42″N/74°50′26″W. Note the subvertical joints. The scale is divided in 0.5-m intervals.

base, probably in an open-shelf to gentle-slope environment. There was sufficient circulation and oxygenation of water, allowing the establishment of moderately diverse marine invertebrate communities. Coarse-grained siltstone (approaching very fine grained sandstone) is more common in the Mahantango in Pike County than it is in Monroe County to the southwest. This feature, coupled with the increasing thickness of the Mahantango towards the eastern corner of Pike County, may reflect proximity to a major Middle-Late Devonian sediment-dispersal center, probably in the vicinity of Port Jervis, New York. Sedimentation rates in this area were apparently high enough to preclude the development of biostromes or laterally persistent and mappable fossiliferous siltstones similar to those of Monroe and Carbon Counties to the southwest.

## TRIMMERS ROCK FORMATION

INTRODUCTION: The Trimmers Rock Formation has been mapped in northeastern Pennsylvania from Carbon County through Monroe County and into southern Pike County (Alvord and Drake, 1971; Bucek, 1971; Schultz, 1973; Epstein and others, 1974; Berg, 1975; Sevon, 1975a; and Berg and others, 1977). From the Birchwood Lakes area (southeastern Edgemere quadrangle) to Millrift (on the Delaware River), Fletcher and Woodrow (1970) mapped the succession from the top of their Sparrow Bush Formation (Tully correlative) to the base of their Delaware River Formation as the Sloat Brook and Millrift Formations (Figure 3). The Millrift overlies the Sloat Brook, and the two stratigraphic units are correlative with the Trimmers Rock Formation as mapped in southern Pike County. Schultz (1973) divided the Trimmers Rock into informal lower and upper members, indicating that they were equivalent to the Sloat Brook and Millrift. It is the authors' view that the Trimmers Rock Formation is divisible and mappable on the basis of textural contrast. The names Sloat Brook and Millrift are utilized in this report, but they are considered only of member rank.

FIELD OCCURRENCE: The Trimmers Rock Formation is situated in a long, relatively narrow belt approximately 1 to 2 miles northwest of the Delaware River, along the southeast side of the county (Plate 1). The gently northwest-dipping, interbedded sandstone, siltstone, and shale yield a topography that typically includes low, subparallel ledges; these are rounded and subdued somewhat by surficial deposits (especially till). The bedrock ledges are more conspicuous in the Millrift Member than in the Sloat Brook Member because of the greater amount of sandstone in the Millrift. Joint development, generally low bedding dips, and modification by glaciation have all combined to produce a drainage pattern that is near to trellis or angulate, but subdued to the point of being fairly complex.

DESCRIPTION: Siltstone is the dominant lithology in the Trimmers Rock Formation in Pike County. However, the siltstone beds are commonly interbedded with silt shale and sandstone, and, to a lesser degree, with clay shale or claystone. Some exposures, particularly in the Millrift Member, are dominantly sandstone with some siltstone interbeds. The Millrift Member includes approximately 60 percent sandstone and 40 percent siltstone and shale (Fletcher and Woodrow, 1970, p. 12–13). Siltstone and silt shale are more common in the Sloat Brook Member, constituting about 90 percent of the member; the remainder is sandstone. In a few places in the Sloat Brook, very dark gray to grayish-black shale interbeds occur.

Trimmers Rock siltstone and shale are dark gray (N3) to medium dark gray (N4), and the sandstone is commonly medium dark gray (N4). Weathering colors are close to olive gray (5Y3/2), in places with a bluish tinge.

The siltstone and sandstone in this formation are generally nonfissile and thin to medium bedded, although some sandstone in the Millrift Member is thick bedded. The shale is fissile to subfissile and in most places thickly laminated.

All of the Trimmers Rock lithologies appear to have secondary, fracture-type porosity. Permeability may be higher in the siltstone and sandstone because joints are more pervasive and better developed than in the shale. Field observation indicates that joints in the shale tend to be discontinuous and curved and have little or no secondary mineralization. Joints in the siltstone and sandstone commonly have secondary drusy quartz crusts.

Planar, parallel bedding of all scales dominates most of the Trimmers Rock Formation exposures (Figure 9). Ripple bedding is fairly common, especially in the thinly bedded silt shale (Figure 10). In plan view most of the ripples are sinuous, but some interference patterns do occur (Figure 11). Other bedding patterns, which are describable as "wavy" or "irregular," are also fairly common. In the medium- to thick-bedded sandstone and coarse-grained siltstone units, load casts are common (Figure 12). Zones of load casts are laterally persistent. Load casts are more common in the Millrift Member, but they do occur in the Sloat Brook Member as well. Groove casts on bedding surfaces are rare. Crossbedding is present in some coarse-grained siltstone and sandstone beds and tends to be low angle and low amplitude, approaching the "hummocky" crossbedding that is typical of offshore facies (Harms and others, 1975, p. 88).

Cyclicity is common in both members of the Trimmers Rock Formation, and variations of the Bouma cyclic sequence are present. The five divisions of a Bouma sequence include the following, in ascending order, (A) massive or graded sand or coarse silt; (B) parallel-laminated sand or coarse silt; (C) ripple- or convolute-bedded silt; (D) parallel-laminated silt or mud; and (E) parallel-laminated or structureless clay or

Figure 9. Moderately dipping, interbedded siltstone and silt shale of the Sloat Brook Member of the Trimmers Rock Formation exposed in a roadcut along Interstate Route 84, 1.25 miles northeast of Milford at 41°20′41″N/74°47′16″W. The planar, parallel bedding dips northwest (left to right); the prominent joints dip southeast (right to left). The scale is divided in 0.5-m intervals.

Figure 10. Very thinly bedded silt shale in the lowermost part of the Sloat Brook Member of the Trimmers Rock Formation showing typical ripple bedding in cross section, especially to the left of the scale. The exposure is on the west side of a secondary road about 2.3 miles northwest of Dingmans Ferry at 41°14′51″N/74°53′49″W. The scale is divided in inches and centimeters.

mud. Such sequences are attributed to deposition by turbidity currents. Some sequences in the Trimmers Rock lack division A, and divisions D and E are commonly thin. Reversals in grading are seen in some places in division A (Figure 13). In some sequences, division C is very thin, subtle, or lacking. It is possible that Bouma sequences are lacking in parts of the Trimmers Rock, especially in the Millrift Member, because deposition by turbidity currents gave way to shallower, shelf-type deposition, yielding simple interbeds of sandstone and siltstone.

Zones of hard, ellipsoidal nodules occur sporadically in the Trimmers Rock. These nodules are presumed to be similar in composition to nodules in the subjacent Mahantango Formation, which are composed of microcrystalline quartz and chlorite. In a few places, the nodule zones are expressed as broad, thin bands.

Trimmers Rock siltstone disintegrates to hackly fragments and less commonly to platy or chippy fragments. Sandstone in this formation characteristically breaks down to rubbly and slabby pieces and less commonly to blocky pieces. The shale normally disintegrates to chippy to hackly fragments.

The most common invertebrate fossils in the Trimmers Rock are brachiopods, usually concentrated in death assemblages in the bottom divisions of turbidite sequences. At least four different genera occur in the brachiopod concentrations, including spiriferid types. Crinoid columnals and bivalves are also common. Bryozoans are present in the Trimmers Rock, but they are

Figure 11. Interference ripples in siltstone of the Millrift Member of the Trimmers Rock Formation at an outcrop along Dingmans Creek in Childs Park at 41°14′16″N/74°55′01″W (hand provides scale).

uncommon to rare. Few trace fossils occur, and they are only small, obscure burrows. Molluscan faunal zones occur in the Sloat Brook Member. A few bivalves (?*Cimitaria*) occur in the sandstone of the Millrift Member at 41°15′32″N/74°54′16″W in the Edgemere 7.5-minute quadrangle.

MICROSCOPIC PETROLOGY: The siltstone is approximately 20 to 50 percent quartz grains set in a composite groundmass of illite, chlorite, and some microcrystalline silica. Very few grains are feldspar, and few are sedimentary rock fragments. Secondary limonite is finely dispersed in parts of the groundmass, and calcite is rarely present. The sandstone is very fine grained, well-sorted, immature quartzarenite in which the framework comprises mostly single or semicomposite, strongly undulose quartz grains, less than 10 percent rock fragments, and a trace of feldspar grains. The framework is set in a binder, which is predominantly an illite-chlorite matrix; about 10 percent of the binder may be authigenic silica or chert cement. Small amounts of limonite are dispersed in the matrix.

THICKNESS: The Trimmers Rock Formation in Pike County has a maximum thickness of approximately 1,825 feet in the vicinity of Milford. About 5 miles southwest of Milford, the formation thins to approximately 720 feet. Continuing to the southwest, the Trimmers Rock thickens, becoming approximately 1,700 feet thick in the vicinity of Bushkill. Northeast of Milford, where the Delaware River crosses the formation, the thickness decreases to approximately 1,650 feet. In the subsurface, the Trimmers Rock is about 1,535 feet thick in the W. Hess #1 well (Shohola quadrangle), and decreases to 1,135 feet in the Pennsylvania Department of Forests and Waters #1 well (Hawley quadrangle). At the western side of Pike County, in the J. Haag #1 well, the Trimmers Rock is 1,460 feet thick (see Figure 2).

Along the outcrop belt, the Sloat Brook Member has its greatest thickness in the vicinity of Milford, where it is approximately 950 feet thick. The overlying Millrift Member, however, has its greatest thickness in the vicinity of Matamoras, where it is approximately 1,000 feet thick. In all likelihood, the Millrift-Sloat Brook contact drops down section going northeast of Milford, reflecting an earlier influx of coarser sediment in the Matamoras-Port Jervis region during Trimmers Rock deposition.

CONTACTS: The contact between the Trimmers Rock Formation and the underlying Mahantango Formation is difficult to discern in Pike County and is within a gradation that is up to 200 feet thick. The contact is associated with a subtle change in topographic relief. The Trimmers Rock is somewhat more resistant to erosion than the Mahantango, and discontinuous low ridges are more characteristic of the Trimmers Rock. Fletcher and Woodrow (1970, p. 70) described a thin sequence of coquinites, fine-grained sandstone, and siltstone directly above the Mahantango Formation, which they named Sparrow Bush Formation. This unit could not be traced by the present authors any appreciable distance from the Delaware River, and thus it was not utilized to divide the Trimmers Rock and Mahantango. The base of the Trimmers Rock is placed at the first significant appearance of

Figure 12. Large load-cast structures (above scale) in the Trimmers Rock Formation. The exposure of sandstone and siltstone is in a roadcut along Interstate Route 84 approximately 1.1 miles north of Milford at 41°20′37″N/74°48′42″W. The scale is divided in 10-cm intervals.

well-developed graded bedding in Bouma sequences and a preponderance of siltstone.

The top of the Trimmers Rock Formation is placed at the base of the thick, gray, nonmarine sandstone of the Towamensing Member of the Catskill Formation. Trimmers Rock sandstone is poorly crossbedded and thinner than the trough-crossbedded Towamensing sandstone.

The contact between the Sloat Brook and Millrift Members is based on the change from a dominance of siltstone in the Sloat Brook to a dominance of sandstone in the Millrift. This change is apparent on aerial photographs of the outcrop belt; resistant sandstone ridges are more numerous in the Millrift Member.

EXPOSURES: There are abundant exposures of the Trimmers Rock Formation in the outcrop belt northwest of the Delaware River. There are many ledges of northwest-dipping siltstone and sandstone protruding from beneath glacial till and colluvial deposits.

One of the best natural outcrops of the Trimmers Rock (uppermost part of the Sloat Brook Member) is a cliff on the west side of the Delaware River 0.4 mile south of Millrift, at 41°24′12″N/74°44′27″W in the Port Jervis North quadrangle. This outcrop (Figure 14) is called the "elephant feet" by local residents, because erosion along joint planes and bedding has produced rounded columnar shapes that resemble elephant legs.

At the waterfalls above Deep Brook, about 60 feet of sandstone of the Millrift Member is exposed. The section is thinly to thickly bedded, and ripple and "wavy" bedding are common. Load casts are also present. The waterfall section is located at 41°21′55″N/74°46′50″W in

Figure 13. Bouma sequence in the Sloat Brook Member of the Trimmers Rock Formation. The thick, coarse siltstone bed near the bottom of the photograph represents division A and displays reverse grading; brachiopod shell debris occurs near the top of the bed. The thinner, fissile silt shale zones above the thick siltstone bed represent divisions C and/or D. The exposure is at the same location as Figure 9. The scale is divided in 10-cm intervals.

the Milford quadrangle. There are also abundant exposures of the Millrift Member within Childs Park along Dingmans Creek in the vicinity of 41°14′10″N/74°55′00″W in the Lake Maskenozha quadrangle.

Some of the Sloat Brook Member lithologies are well exposed along Interstate Route 84, 1.25 miles northeast of Milford at 41°20′41″N/74°47′16″W in the Milford quadrangle (Figure 9). Exposures of the Sloat Brook Member are also common along a tributary of Toms Creek, beginning in the vicinity of 41°08′44″N/74°58′07″W in the Lake Maskenozha quadrangle. The lower part of the Sloat Brook Member is well exposed in a deep ravine at Winona Falls, at 41°06′33″N/75°02′52″W in the Bushkill quadrangle.

ORIGIN: The entire Trimmers Rock Formation was deposited in a marine prodelta setting which is best characterized as "distal" for the lower part of the formation and "proximal" for the upper part. Bouma sequences and sedimentary structures, including load casts and current ripples, point to deposition by density currents below effective wave base. The sandstone in the Millrift Member was probably deposited in delta-front environments and marginal-marine envi-

Figure 14. "Elephant feet" in the uppermost part of the Sloat Brook Member of the Trimmers Rock Formation along the Delaware River 0.4 mile south of Millrift. The exposure comprises thin siltstone beds and a few thin, very fine grained sandstone beds. Weathering and erosion along the joint planes and bedding planes has produced the broad, rounded columnar shapes. The exposure is located at 41°24′12″N/74°44′27″W.

ronments associated with the beginning of a transition from marine to nonmarine. However, rocks representing specific environments such as estuaries or beaches have not been identified. Further discussion of the stratigraphy and origin of the Trimmers Rock Formation can be found in Schultz (1974).

## CATSKILL FORMATION

### Introduction

The first stratigraphic nomenclature applied to the sequence of nonmarine rocks (Catskill) in Pike County was that of White in 1882 (Figure 3). White recognized a relatively simple but distinctive subdivision of rocks above the marine rocks he called Chemung (Trimmers Rock). He applied the previously used names, Starrucca sandstone, New Milford red shale, and Montrose red shale of adjacent Wayne County to some of the sequence and introduced a new name, Delaware river flags, to replace the names Paupack sandstone and New Milford sandstone, which he had used in Wayne County (White, 1881). Prosser (1894) retained the nomenclature used by White and changed only the biostratigraphic position of the Chemung. He did not, as suggested by Fletcher and Woodrow (1970, p. 15), apply the name Delaware river flags to a sequence of rocks other than those indicated by White.

Willard and others (1939) completely changed the stratigraphic terminology for northeastern Pennsylvania from that of White. This was apparently the result of Willard's emphasis on biostratigraphy rather than lithostratigraphy. Fletcher and Woodrow (1970) also changed the stratigraphic nomenclature for northeastern Pennsylvania by revising the definition of Willard's Shohola, defining a new unit (Lackawaxen), and introducing some terminology from New York State (Walton).

The stratigraphic units shown on the 1970 Geologic Map of New York (Fisher and others, 1970) cannot be easily rectified with any of the mappable rock units recognized in this report. The Slide Mountain Formation apparently correlates with at least the base of the Lackawaxen Member. The overlying Honesdale Formation, which derives its name from Honesdale, Pennsylvania, is not considered a mappable rock unit in Pennsylvania, and its correlation (at Honesdale) with the base of the Poplar Gap and Packerton Members (undivided) was established during reconnaissance mapping for the 1980 Geologic Map of Pennsylvania (Berg and others, 1980).

Stratigraphic subdivisions of the Catskill Formation in adjacent areas of Pennsylvania are based on stratigraphy established along the Lehigh River in Carbon County (Epstein and others, 1974) about 40 miles from the southwest border of Pike County and about 65 miles from the good exposures along the Delaware River just northwest of Matamoras. Because of the considerable differences in facies that resulted from deposition of the Catskill by several different sediment-input systems (Sevon, 1979; Thompson and Sevon, 1982), the authors believe that use of local and/or original names (White, 1882) is desirable in Pike County. Therefore, established names have been used wherever appropriate in order to clarify previous use and correlation.

## Towamensing Member

INTRODUCTION: In Susquehanna and Wayne Counties, White (1881, p. 70) recognized a thin interval of olive-colored shale and sandstone which occurred below the first red beds of the Catskill and above the rocks he called Chemung (essentially the Trimmers Rock Formation of this report). He called this interval the Starrucca beds. In Pike County, he recognized a similar but considerably thicker unit in the same stratigraphic position and applied the same name (White, 1882, p. 102; Figure 3). He stated that the interval was "finely exposed along the New York side of the Delaware river between Pond Eddy, and one half mile above the Erie R. R. bridge across the Delaware . . . ." This exposure is known today as Hawks Nest. White (1882, p. 80, 103) traced the unit across Pike, Monroe, and Carbon Counties to excellent exposures on the Lehigh River, but did not apply the name Starrucca to these exposures. He identified the top of the Starrucca as the base of the first red shale in the Catskill and the base of the Starrucca as occurring where "the hard greenish-gray, micaceous Catskill sandstones are underlaid by blue, sandy, fossiliferous shales; and these by olive and gray micaceous sandstones which are destitute of the peculiar greenish cast so common in the Catskill" (1882, p. 104).

Prosser (1894) used Starrucca in the same context as White. Willard (1936, p. 587–588), having apparent disregard for White's definitions and clear locations of specific rocks, applied the name Delaware River flags (see the following section "Delaware River Member") to the interval White called Starrucca. Willard made no mention of the term Starrucca in 1936, but commented in 1939 (Willard and others, p. 299) that the term Starrucca as used in Wayne and Susquehanna Counties had no basis for retention. Fletcher and Woodrow (1970, p. 15) applied the name Delaware River Formation to the interval called Starrucca by White (see the section "Delaware River Member"). Alvord and Drake (1971) applied the name Delaware River Flags Member to this stratigraphic interval in southwestern Pike County and northeastern Monroe County.

Mapping for this report and elsewhere to the southwest of Pike County (Bucek, 1971; Epstein and others, 1974; Berg, 1975; Sevon, 1975a, b, c; Berg and others, 1977) indicates that the interval that White named Starrucca in Pike County is persistent from the Delaware River to the Lehigh River (and farther southwest) and that White's correlations of the Starrucca (1882, p. 75–80, 103) are correct.

The name Towamensing Member was applied to this stratigraphic interval in the Lehigh River area by Epstein and others (1974, p. 158–162) and has been utilized to the northeast in Carbon and Monroe Counties by Berg (1975), Sevon (1975a), and Berg and others (1977) and to the southwest in Carbon and Schuylkill Counties by Wood (1974a, b). The name Towamensing Member is used in this report because (1) the name has widespread usage elsewhere in northeastern Pennsylvania; (2) the unit has proven continuity from the Towamensing type locality on the Lehigh River to the Delaware River; and (3) the interval in Pike County that was called Starrucca by White cannot be correlated directly with his type Starrucca, although stratigraphic position argues for the correctness of the correlation. In addition, reconnaissance in the Starrucca area (Wayne County) indicates that the rocks that White called Starrucca in that area are physically somewhat dissimilar to those he called Starrucca in Pike County, whereas there is considerable similarity of the stratigraphic interval everywhere between Hawks Nest and the Lehigh River.

FIELD OCCURRENCE: The Towamensing Member occurs as a band of gray sandstone and dark-gray shale across southeastern Pike County from the Delaware River to Monroe County (see Plate 1). Ledges of sandstone occur throughout the outcrop belt, but glaciation has smoothed much of the area underlain by this relatively uniform rock sequence, and low, rounded hills are the characteristic landform; outcrops are sparse in much of the county. Several small streams have cut valleys across the outcrop belt, but drainage is poorly developed in some places between these streams.

DESCRIPTION: The Towamensing Member comprises mainly sandstone, some interbedded siltstone and claystone, and some intraformational conglomerate. Carbonized plant fragments and burrows of the clam *Archanodon catskillensis* (Vanuxem) have been found in the unit.

Sandstone in the Towamensing Member is mainly light to medium gray, having a bluish tint in some places and an olive or greenish tint in other places. Grain size ranges from very fine to medium, and most of the sandstone is fine to medium grained. No coarse-grained sandstone or conglomeratic beds have been observed in this member in Pike County. Most of the sandstone is crossbedded in broad troughs up to several feet in width (Figure 15). Bed sets are up to 6 feet thick, but in most places are 3 feet thick or less. Crossbed inclination varies from steep to shallow. Planar-bedded sandstone occurs (Figure 16), but plane bedding is not as common as trough crossbedding and seems to be best developed near the Delaware River northwest of Matamoras. Linguoid ripples and parting-step lineations are common on bedding surfaces, particularly in the upper parts of planar-bedded sequences (Figure 17). Sequences vary from fining upward to coarsening upward. Planar-bedded sandstone is overlain by siltstone and claystone

Figure 15. Crossbedded sandstone of the Towamensing Member exposed in the Eureka Stone Company quarry about 3.5 miles northwest of the center of Milford on the south side of U.S. Route 6 at 41°20′50″N/74°51′43″W.

Figure 16. Planar-bedded and trough-crossbedded sandstones of the Towamensing Member exposed in an old flagstone quarry 2.6 miles southeast of Rosas at 41°24′44″N/74°46′39″W. The person is 5 feet tall.

in some places and by coarser grained crossbedded sandstone in other places. Carbonized plant fragments occur on some bedding surfaces of the sandstone. These fragments are generally 1 to 2 cm in length, but pieces up to 60 cm in length have been observed. Large, vertical, tubelike sedimentary structures occur in the sandstone in the lower part of the Towamensing Member (Figure 18). These structures have been interpreted (Berg, 1977; Thoms and Berg, 1985) to be escape burrows of the freshwater clam *Archanodon catskillensis* (Vanuxem), and are apparently restricted to the sedimentary facies represented by the lower part of the Towamensing.

Initial breakup of the sandstone is controlled by joints and bed sets and produces blocks, which further disintegrate into slabby and platy pieces.

Numerous thin, interbedded siltstones and claystones occur in the Towamensing Member. Most of these beds are medium dark to dark gray, particularly in the lower part of the unit, but light-olive-gray siltstones and claystones occur in the upper part of the member (Figure 16). The interbeds are generally less than 2 feet thick and pinch out laterally within a few feet to a few tens of feet (Figure 19). None of these finer grained beds appear to have great lateral continuity, although in many areas a limited outcrop size obscures their true extent. Most of the siltstones and claystones have sharp lower and upper contacts, and the upper contact commonly is erosive. The siltstones and claystones range from fissile to structureless and break up into small, hackly or platy pieces. Fragments of carbonized plant material are common on the bedding surfaces of these fine-grained rocks.

Intraformational conglomerates are relatively common in the Towamensing Member. They are lensoid concentrations of dark-gray shale clasts, light-gray calcareous nodules, and plant fragments in a calcite-cemented sand matrix. Quartz

Figure 17. Linguoid ripples (upper) and parting-step lineation (lower) on bedding surfaces of Towamensing Member sandstone exposed on the south bank of the Delaware River about 1.9 miles southeast of Rosas at 41°25′25″N/74°46′49″W. The pens are 5.2 inches long and are parallel to the ripple troughs and to the parting lineation.

Figure 18. Vertical to subvertical, tubelike structures in Towamensing Member sandstone exposed at Hawks Nest on the east side of N.Y. Route 97 at about 41°25′00″N/74°44′07″W, 1.3 miles northwest of Sparrow Bush, New York. The structures, interpreted as escape burrows of the freshwater bivalve *Archanodon catskillensis* (Vanuxem), are just a few of many exposed in the southern part of the Hawks Nest roadcut.

pebbles up to 1 cm in diameter occur rarely, as do fragments of fish plates. The shale clasts are platy in shape, have rounded edges, and are up to 8 cm, but generally less than 4 cm, in diameter. The calcareous nodules are either rounded masses up to 8 cm in diameter containing concentric internal structures (pisoliths), or they are rounded, rod-shaped pieces less than 1 cm in diameter and 2 to 3 cm long. The nodules are not always present in the conglomerates and are not nearly as abundant as they are in higher stratigraphic units. The intraformational conglomerates have relatively sharp boundaries and are generally less than 2 feet thick and 3 to 6 feet long. The lenses generally occur at the base of crossbedded sandstone overlying an erosive basal contact. The structure within the conglomerates is usually chaotic, but in some places crossbed laminae persist from adjacent sandstone through lensoid accumulations. Where this occurs the conglomerates commonly contain a relatively small amount of clast-sized material.

The intraformational conglomerates weather rapidly compared with the adjacent noncalcareous sandstone. Initial weathering results in the development of a pale-brown (5YR5/2) color on the exposed surface, which contrasts markedly with the surrounding light- to medium-gray colors. The pale-brown color defines clearly the boundaries of the conglomerates and is caused by the weathering of the calcite cement. Further weathering results in disintegration of the rock, gradual pitting of the surface, and change to a

Figure 19. Crossbedded and planar-bedded sandstones and a siltstone interbed (outlined) of the Towamensing Member exposed along the Erie-Lackawanna Railroad on the south bank of the Delaware River about 1.3 miles north-northwest of Millrift at 41°25′39″N/74°45′00″W.

dark-brown to black color. In deeply weathered exposures the conglomerates may be recessed several to many inches and no fresh rock can be found. Such deeply weathered zones are crumbly and relatively easy to excavate, although they may be rather ‘‘punky’’ when struck by a hammer.

MICROSCOPIC PETROLOGY: Point-count analyses of 22 thin sections of sandstone and two thin sections of intraformational conglomerate from the Towamensing Member are summarized in Table 1.

The dominant framework component in the sandstone is moderately to highly strained, subangular quartz having moderate to large amounts of secondary silica overgrowth. Shale fragments are the next most abundant framework component. Feldspar is commonly present, but in small amounts. Heavy minerals are sparse and are mainly zircon, although apatite, magnetite, and leucoxene do occur. Slightly diffuse masses of limonite make up about 1 percent of most of the samples. Moderate chloritization of illite has occurred in almost all of the samples. Most of the samples from planar-bedded sandstone have a good alignment of the long axes of quartz grains and of mica flakes parallel to laminae. The average grain size for all of the sandstone thin sections is 0.13 mm.

Only two thin sections of intraformational conglomerate were examined, and they are almost equally divided between framework components and cement. Strained quartz grains and carbonate masses make up equal amounts of the framework (20 percent each), and unstrained quartz (present in one thin section only) and shale fragments make up most of the remainder. Sparry calcite is the main cement and totally surrounds most of the framework grains. Illite-chlorite-chert

Table 1. *Microscopic Composition of the Towamensing Member (24 Samples)*

(Values represent percentage of total composition)

| Component | Sandstone (22 samples) | Intraformational conglomerate (2 samples) |
|---|---|---|
| Framework | 64 | 50 |
| Cement | 35 | 49 |
| Matrix | 1 | 1 |
| Framework | | |
| Strained quartz | 53 | 20 |
| Unstrained quartz | 0 | 5 |
| Feldspar | 1 | Tr[1] |
| Microcrystalline quartz | Tr | Tr |
| Polycrystalline quartz | 1 | Tr |
| Stretched metaquartz | Tr | Tr |
| Shale clasts | 5 | 4 |
| Siltstone clasts | Tr | Tr |
| Metamorphic rock fragments | 1 | Tr |
| Coarse micas | 2 | Tr |
| Carbonate clasts | 0 | 20 |
| Miscellaneous | 1 | 1 |
| Cement | | |
| Illite-chlorite-chert composite | 23 | 9 |
| Microcrystalline silica | 4 | 1 |
| Authigenic silica overgrowths | 8 | 4 |
| Calcite | Tr | 35 |
| Matrix | | |
| Kaolinite, illite, chlorite, and limonite | 1 | 1 |

[1]Tr, trace.

cement is present in small pockets surrounded by clusters of larger grains.

One thin section of claystone (not in Table 1) shows that 84 percent of the rock is well-chloritized illite, 10 percent microcrystalline silica, and 6 percent angular, strained quartz grains 0.01 to 0.02 mm in diameter. The illite flakes are well aligned.

THICKNESS: The thickness of the Towamensing Member varies considerably in Pike County; the unit has a pronounced thinning away from its maximum in the eastern corner of the state (Figure 20). The maximum calculated thickness, near the Delaware River in the northeastern corner of the county, is 1,625 feet; the thinnest interval is 247 feet, in the western corner of the county.

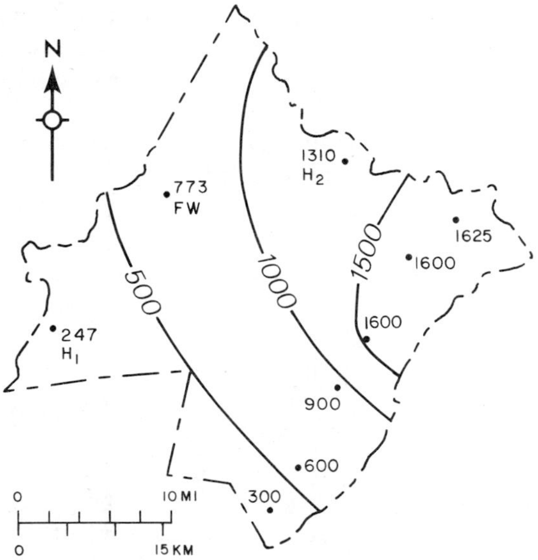


Figure 20. Isopach map showing the thickness of the Towamensing Member in Pike County. The control thicknesses in the southeast part of the county were calculated from the geologic map (Plate 1). Subsurface control was derived from the well data presented in Figure 2: $H_1$, J. Haag #1; $H_2$, W. Hess #1; and FW, Pennsylvania Department of Forests and Waters #1.

CONTACTS: The base of the Towamensing Member is placed at the base of the lowest thick, gray, nonmarine sandstone that occurs above the interbedded sandstone, siltstone, and shale of the Trimmers Rock Formation. Although placement of the actual contact may be arbitrary in areas of poor exposure, in most exposures the differences between the Towamensing and Trimmers Rock allow fairly accurate placement. The upper contact is placed at the base of the first red-bed unit in the Catskill Formation, the base of the Delaware River Member.

EXPOSURES: The best exposure of the Towamensing Member occurs along the east side of N.Y. Route 97, 1.3 miles northwest of Sparrow Bush, New York, in what is known as the Hawks Nest section. This superb roadcut, which is about 1 mile long, exposes most of the member and has the best display of *Archanodon* burrows (see Figure 18). Additional exposures occur along the bank of the Delaware River and the old Delaware Canal below the road level. One of the best exposures within Pike County is in the Eureka Stone Company quarry northwest of Milford (see Figure 15 for the exact location). Very good exposures also occur along both lanes of Interstate Route 84 at 41°20′15″N/74°52′01″W about 1.75 miles west (road distance) of the intersection of Interstate Route 84 and U.S. Route 6. Some good exposures occur in old flagstone quarries near the Delaware River (e.g., Figure 16), but these are not readily accessible. Small, scattered exposures occur throughout the outcrop belt in Pike County, but none show a significant amount of the member.

ORIGIN: The Towamensing Member contains the oldest sediments of nonmarine origin in the Catskill Formation in Pike County. The overall character is fluvial, but the unit lacks the red beds indicative of extended periods of subaerial exposure typical of stratigraphically higher rocks of the Catskill. The Towamensing may be best explained as the result of deposition on a fluvial-dominated delta plain, which received abundant sediment supply and subsided rapidly. Deposition was by shallow, low-sinuosity streams, probably braided. The mixture of coarsening-upward and fining-upward sequences, the thin, interbedded siltstone and claystone, the common occurrence of shale clasts and plant material, and the apparent lack of soil development are strikingly similar to characteristics of deposits described by Steel and Aasheim (1978).

The transition from marine to nonmarine environments of deposition presumably occurs at the base of the Towamensing or at the top of the underlying Trimmers Rock Formation, but the exact nature of that transition and its associated rocks is not well understood, perhaps because of poor exposure of this interval, but more likely because of a lack of sufficiently detailed study. The presence of the *Archanodon* burrow structures at the base of the unit indicates that the Towamensing is nonmarine, but presumably close to the marine-nonmarine interface (Berg, 1977; Thoms and Berg, 1985). The lower part of the Towamensing, which contains the burrows, is possibly the result of deposition in river-mouth bars. Clearly defined shoreface deposits are typically absent from the Late Devonian sequence in the eastern outcrop belt in Pennsylvania, and tidal-channel deposits have not been described, although they may occur in the underlying Trimmers Rock. Additional detailed work on the sedimentology of this interval is required in order to clearly define the marine-nonmarine interface.

MISCELLANEOUS: The planar-bedded sandstone in the upper part of the Towamensing Member has been quarried to some extent for flagstone (see the section "Bluestone (Flagstone)," on p. 113–117). Excellent-quality crushed rock is produced from the member at the Eureka Stone Company quarry, and the potential for similar crushed-rock production occurs throughout the outcrop belt.

## Delaware River Member

INTRODUCTION: The Delaware River flags were originally defined by White (1882, p. 99) as "the series of greenish gray sandstone and shale which intervene between the base of the Montrose red shale and the top of the New Milford red shale" (see Figure 3). He went on to say that "nearly all the great flagstone quarries along the Delaware are found at various horizons in these measures" and "the rocks composing this series are finely exposed along the Delaware river between Narrowsburg and Pond Eddy, and along this line have a thickness of 1430′ . . . ." White recognized this sequence of rocks as a sand-dominated unit lying immediately above the first red beds in the Catskill series. He correlated the unit across Pike, Monroe, and Carbon Counties to the Lehigh River. Although his correlation maintained a correct stratigraphic order, it is clear from his section descriptions for Monroe County and the Lehigh River (White, 1882, p. 77–80) that he correlated the Delaware River flags with what is now mapped as the Beaverdam Run Member of the Catskill Formation (Epstein and others, 1974, p. 165–171, Plate 3). White (1882, p. 100) included the Lackawaxen conglomerate within the sequence and noted that the conglomerate, which he considered to be a single bed, occurs near the top of the Delaware River flags. However, he did not clearly define a top for the flags.

Prosser (1894) used the name Delaware River flags in the same context as White and made the same mistake in correlation for Monroe County and the Lehigh River. Willard having apparent disregard for White's definition, applied the name Delaware River flags to the interval that White called Starrucca (White, 1882, p. 102) and used a new name, Shohola Formation, for the sequence that White called Delaware River flags (Willard, 1936, 585–588). It is difficult to fully understand why Willard misapplied the name Delaware River flags in Pike County, but some insights are gained from a later discussion of the unit (Willard and others, 1939, p. 295). Willard recognized that the unit on the Lehigh River that White called Delaware River flags (now mapped as the Beaverdam Run Member) was marine in origin and that farther west the unit became difficult to distinguish from the stratigraphically lower Trimmers Rock Formation (the Chemung of White). Willard also assumed that the red shales underlying the Delaware River flags of

White in Monroe County were local in occurrence. Thus Willard probably thought that the Delaware River flags belonged immediately above the marine sequence in Pike County and that White had misidentified his units. Willard alludes to this in his discussion of the Shohola Formation (1936, p. 587).

Fletcher and Woodrow (1970, p. 15) applied the name Delaware River Formation to the approximate interval originally called Starrucca by White. They suggested that the downward shift of the name stemmed from Prosser (1894), but careful reading of Prosser reveals no place where he altered the definition of White's stratigraphic units. In fact, Prosser (p. 43–44) specifically referred to the Hawks Nest exposure, designated by Fletcher and Woodrow (p. 16) as the type locality of their Delaware River Formation, and called it Starrucca. In addition, Prosser (p. 78) included a summary stratigraphic column in which he used White's original terminology. For the interval designated as the Delaware River flags by White, Fletcher and Woodrow (p. 18–19) accepted and used the name Shohola Formation introduced by Willard.

Field work for this report has demonstrated that a mappable stratigraphic unit exists in Pike County that is essentially synonymous with the unit that White called the Delaware River flags. Therefore, the authors retain most of the name and formalize it as the Delaware River Member of the Catskill Formation. They also reject the application of the name Delaware River to other stratigraphic intervals as was done by Willard (1936), Fletcher and Woodrow (1970), and Alvord and Drake (1971). The base of this member is somewhat lower than that identified by White, and the member includes the red beds called New Milford red shale by White (1882, p. 101). The top of the member is placed at the base of the overlying Lackawaxen Member or the undivided Poplar Gap and Packerton Members, both of which are readily mappable units. Both of the contacts are in keeping with the spirit of the Delaware River flags as defined by White: he recognized the New Milford as the lowest red beds in the sequence, and field mapping for this report has merely located even lower red-bed occurrences; he indicated that the Lackawaxen conglomerate was near the top of the sequence, and the authors use it (where present) as the base of the overlying unit (Lackawaxen Member) because of its mappability.

The Delaware River Member correlates directly with (from youngest to oldest) the Long Run, Beaverdam Run, and Walcksville Members of the Catskill Formation in Monroe and Carbon Counties (Bucek, 1971; Epstein and others, 1974; Berg, 1975; Sevon, 1975a; Berg and others, 1977; Sevon and Berg, 1978). The Beaverdam Run Member pinches out through facies change about 8 miles southwest of Pike County, and the adjacent members are combined as an individual unit on the 1980 Geologic Map of Pennsylvania (Berg and others, 1980). Because of the distinctiveness of the Delaware River Member and the facies changes that occur within the member to the southwest, the authors recommend that the name not be used southwest of the Pike-Monroe county boundary.

Fletcher and Woodrow (1970, p. 18) indicated that the Delaware River Member of this report correlates with the lower part of the Walton Formation in New York State. The validity of their correlation has not been verified for this report, nor has the New York terminology been used.

FIELD OCCURRENCE: The Delaware River Member occurs as a broad band of dominantly gray sandstone and some red siltstone and claystone across Pike County from the Delaware River to Monroe County (see Plate 1). Because of the dominance of sandstone in the unit, ledges are the characteristic landform. These are particularly well developed on the steep slopes along the Delaware River and its tributary valleys. Southwest of the Delaware River, where steep slopes are generally lacking and glacial deposits are more abundant, ledges are less common, and in some areas outcrops are sparse.

DESCRIPTION: Most of the Delaware River Member is a gray, planar-bedded or crossbedded sandstone. Some thin, red siltstone and claystone occur in the northeastern part of the county, and thicker, more abundant, red sequences occur in the southwestern part of the county. Thin conglomeratic zones occur locally near the top of the member.

The Delaware River Member in the northeastern part of the county contains repeated cyclic sequences, which are represented in Figure 21. All of the components of the idealized sequence are not always present, but their order appears to be very consistent. The base of the sequence is rarely exposed, but is presumably always marked by a sharp contact with the underlying rocks.

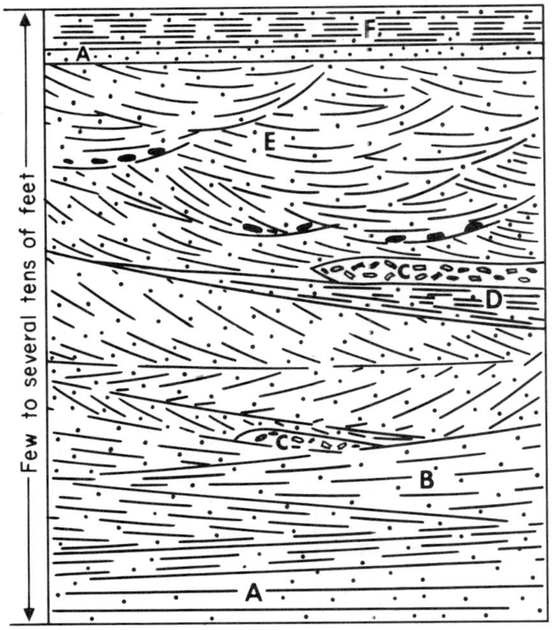


Figure 21. Idealized cyclic sequence typical of the Delaware River Member in northeastern Pike County. Within the sequence there is an upward increase in average grain size and strata thickness and a decrease of crossbed wavelength. A, planar-bedded sandstone; B, broad trough-crossbedded sandstone; C, intraformational conglomerate; D, dark- or light-olive-gray siltstone or claystone; E, small- to medium-scale, trough-crossbedded sandstone; and F, red siltstone or claystone.

Above the base of the sequence is planar-bedded sandstone, which is up to a few feet thick (Figure 22). The sandstone is thinly laminated in most places, but tends to part along planes separated by an inch or more. Bedding surfaces commonly have parting and parting-step lineations. Many bedding surfaces have randomly scattered plant fragments generally less than 2 cm in length. Most laminations are less than 0.5 cm thick and very uniform, but reactivation surfaces having a relief of 1 to 2 cm do occur. Sandstone texture ranges from very fine to medium grained, but fine to medium grained is the most common.

The planar-bedded sandstone is overlain, either gradationally (Figure 23) or sharply (Figure 22), by a slightly coarser grained sandstone having broad trough crossbeds. The troughs are commonly many feet wide, and bedding may appear nearly planar. The character of the trough-crossbedded sandstone is essentially the same as that of the underlying planar-bedded sandstone.

Figure 22. Planar-bedded sandstone typical of the Delaware River Member in northeastern Pike County. The exposure is in an old flagstone quarry along the south side of a pipeline about 1 mile west of the settlement of Walker Lake at 41°25′13″N/74°56′29″W. Note the sharp contact with the overlying, poorly defined beds in the upper third of the exposure.

Figure 23. Gradational upward change from planar-bedded to broad, trough-crossbedded sandstone in the Delaware River Member in northeastern Pike County. The exposure is in an old flagstone quarry 0.4 mile southeast of the south end of a bridge across the Delaware River between Pond Eddy, New York, and Pond Eddy, Pennsylvania, at 41°26′06″N/74°48′47″W. The hammer (right center of vertical rock face) is 11 inches long.

Figure 24. Interbedded sandstone and dark-gray siltstone in the Delaware River Member exposed on the east bank of the Delaware River 0.7 mile southeast of Masthope at 41°31′56″N/75°00′59″W.

In some exposures, bedding is very poorly defined, and there may be one or more interbeds of medium-dark-gray (N4) to dark-gray (N3) or light-olive-gray (5Y5/2) siltstone and/or silty claystone (Figure 24). The interbeds are commonly less than 1.5 feet thick, lensoid, and fissile to subfissile. They contain carbonized plant material in some places, and may have bedding disorientation in thin upper parts. Areal dimensions of individual interbeds are usually indeterminate, but the dimensions of most interbeds probably are less than a few tens of feet.

In addition to the siltstone-claystone interbeds, the interval of broad trough-crossbedded sandstone may also contain thin, intraformational conglomerates (Figure 25). The intraformational conglomerates comprise a chaotic mixture of dark siltstone clasts, calcrete nodules, some quartz pebbles, and carbonized plant material enclosed

Figure 25. Sequence of planar-bedded sandstone (base), intraformational conglomerate (recessed zone), irregularly bedded to broad, trough-crossbedded sandstone (middle), and small- to medium-scale, trough-crossbedded sandstone (top) in the Delaware River Member exposed on the east bank of the Delaware River 0.7 mile southeast of Masthope at 41°31′56″N/75°00′59″W.

in a sand matrix and cemented with calcite. The conglomerates occur at the base of sandstone beds, overlie a sharp, erosive contact, are thin to discontinuous or lensoid, and weather more rapidly than adjacent sandstone.

The upper part of the cyclic sequence is composed of sandstone characterized by medium- to small-scale, trough crossbedding (Figures 25 and 26). The troughs are generally less than 9 feet wide and 1.5 feet thick. Numerous troughs compose single bed sets generally less than 3 feet thick and in some places less than 1.5 feet thick. Bed sets have sharp, erosive bases, and the interval of medium- to small-scale, trough-crossbedded sandstone is commonly multistoried. The sandstone ranges from fine to coarse grained, but most is medium grained. Small, rounded quartz pebbles up to a centimeter in diameter occur in some places, as do dark-gray, siltstone-claystone clasts. The siltstone-claystone clasts are flat shaped and generally less than 2.5 cm long, but may be as long as 10 cm. The quartz pebbles may occur anywhere within a crossbed trough, but the siltstone-claystone clasts commonly occur near the base of a trough. Intraformational conglomerates like those previously described occur in a few places within the interval and are located at the bases of troughs.

The top of the cyclic sequence may have a thin layer of very fine grained, gray- or grayish-red sandstone, which grades upward into red siltstone and claystone. The red siltstone and claystone have root casts, burrows, and desiccation cracks (Figure 27). The red-bed units are exposed in only a few places and probably do not occur at the tops of most of the cyclic sequences, which are repeated many times in the Delaware River Member.

There is a gradual change of facies in the Delaware River Member southwestward across the county. The amount of planar-bedded sandstone decreases and the amount of red siltstone and claystone increases. Trough crossbedding, both broad and medium to small scale, is the dominant form of bedding in the sandstone, but in many small exposures, bedding is obscure and difficult to interpret. Bed sets are commonly less than 3 feet thick, and contacts are always sharp

Figure 26. Small-scale, trough-crossbedded sandstone typical of the upper part of a cyclic sequence in the Delaware River Member. The exposure is on the east bank of the Delaware River directly opposite Masthope at 41°32′15″N/75°01′29″W. The hammer handle points down the trough axis, which is oriented N18°W.

Figure 27. Desiccation cracks in grayish-red claystone of the Delaware River Member exposed along the south bank of the Delaware River about 0.6 mile west of the mouth of Pond Eddy Creek at 41°25′52″N/75°50′16″W. The pen is 5.2 inches long.

and erosional. Thin to moderately thick intervals of red siltstone and claystone are common and crop out in many places. These fine-grained rocks contain root casts, burrows, mud cracks, and dipnoan lungfish aestivation burrows (Figures 28 and 29). A sequence typical of the Delaware River Member in western Pike County is shown in Figure 30. The sandstone in this member tends to be coarser in western Pike County, and coarse-grained sandstone beds are common. Scattered quartz pebbles, thin (less than a few inches) conglomeratic beds, and lenses of intraformational conglomerate are also common. Compositional differences occur and are discussed under "Microscopic Petrology."

Discontinuous, thin conglomeratic zones occur near the top of the member in both the northeastern and southwestern parts of the county, but appear to be absent in the central part of the county (Plate 1). In the northeastern part of the county a thin conglomeratic sandstone is present about 400 feet below the top of the Delaware River Member and has been traced from Shohola to the vicinity of Twin Lakes. This zone was identified as the base of the Lackawaxen Formation by Fletcher and Woodrow (1970, p. 20). A second conglomeratic sandstone occurs at a higher stratigraphic level, approximately 200 feet below the top of the member, in the area between Shohola Creek and Walker Lake. These conglomeratic zones comprise mainly coarse- to very coarse grained, gray sandstone with 3 to 6 feet of conglomerate in variable concentrations (Fig-

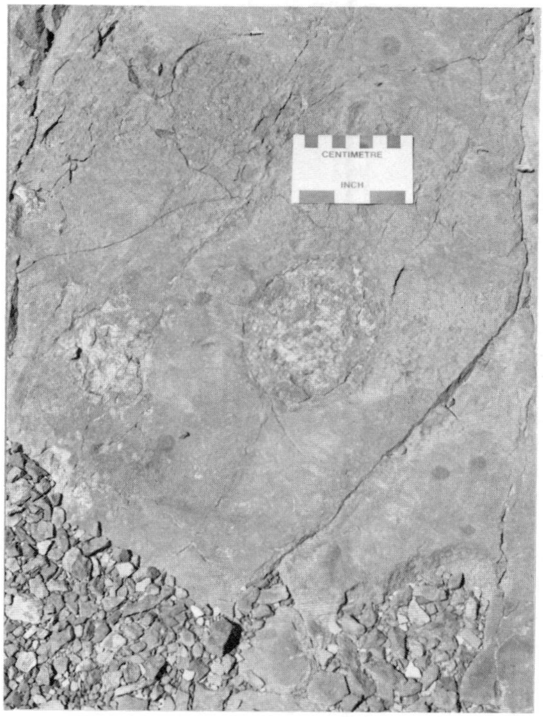


Figure 29. Cross section of a filled aestivation burrow (below scale) of a dipnoan lungfish on the bedding surface of a grayish-red claystone in the Delaware River Member. The exposure is near the escarpment edge of the Pocono Plateau on East Mountain at 41°15′02″N/75°13′04″W. The scale is divided in inches and centimeters.

Figure 28. Vertical, filled aestivation burrow of a dipnoan lungfish in grayish-red siltstone of the Delaware River Member. The exposure is in a small borrow pit about 0.35 mile northwest of Indian Ladder Falls at 41°14′58″N/75°14′22″W. The scale is divided in inches and centimeters.

Figure 30. Sequence of grayish-red siltstone and claystone (lower) and gray sandstone (upper) of the Delaware River Member exposed along the west-bound lane of Interstate Route 84 about 0.88 mile east of the Pike county line at 41°20′36″N/ 75°19′31″W.

ure 31). In some places a bed up to 3 feet thick and containing up to 75 percent pebbles (Figure 32) may be interbedded with coarse-grained sandstone, which either lacks pebbles or contains a few scattered quartz pebbles. The conglomeratic beds are trough crossbedded, but stratification may be obscure. Neither of the conglomeratic zones is traceable for any distance to the southwest, and the lower zone appears to be discontinuous in the area where it has been identified.

In the southwestern part of the county another conglomeratic zone occurs about 60 feet below the top of the member at the edge of the Pocono Plateau escarpment at East Mountain. This conglomeratic zone is apparently very local in occurrence and has not been traced beyond the limits shown on Plate 1. The conglomeratic zone is defined by about 6 feet of very well cemented, medium to very coarse grained, gray sandstone containing scattered pebbles up to 2.5 cm in diameter. At the top of this zone there are local concentrations of pebbles, which constitute up to 50 percent of the rock and are mainly 0.5 to 1.2 cm in diameter. The sandstone is trough crossbedded, but the crossbedding is obscured by cementation. Bed sets are 1.5 to 3 feet thick and have sharp boundaries. The zone is underlain and overlain by similar types of sandstone which generally lack pebbles.

Disintegration of the various lithologies is controlled by spacing of joints and stratification. Well-cemented sandstone and conglomerate initially break up into blocks whose dimensions are determined by bed-set thickness and joint spacing. These blocks subsequently break up into smaller blocks and platy pieces along parting planes between planar or crossbedded laminae sets. The weathering of these rocks by granular disintegration is slow. Only about 5 mm of sand matrix surrounding glacially striated quartz pebbles exposed at the top of a ledge on East Mountain has been removed in approximately 20,000 years (Sevon, 1984).

MICROSCOPIC PETROLOGY: Point-count analyses of 75 thin sections of sandstone, seven thin sections of intraformational conglomerate, and four thin sections of conglomeratic sandstone from the Delaware River Member are summarized in Table 2.

The dominant framework component in the sandstone is moderately to highly strained, subangular quartz having variable amounts of quartz overgrowth. The sandstone in the eastern part of the county contains more feldspar and a greater variety of cement, whereas the sandstone in the western part of the county contains more quartz and shale clasts. Feldspar is generally weathered to some extent and often partially replaced by calcite. Platy minerals and the long axes of quartz grains are aligned parallel to laminae in the planar-bedded sandstone; no preferred orientation is apparent in the crossbedded sandstone. Leucoxene and zircon are common, though not abundant, heavy minerals. Slightly diffuse masses of limonite occur in almost all of the sandstone samples. Chloritization of illite has occurred in almost all of the samples, and most samples are low to moderate in degree of chloritization. The average grain size for all of the sandstone thin sections is 0.2 mm.

The conglomeratic sandstone from the eastern part of the county has an average grain size of

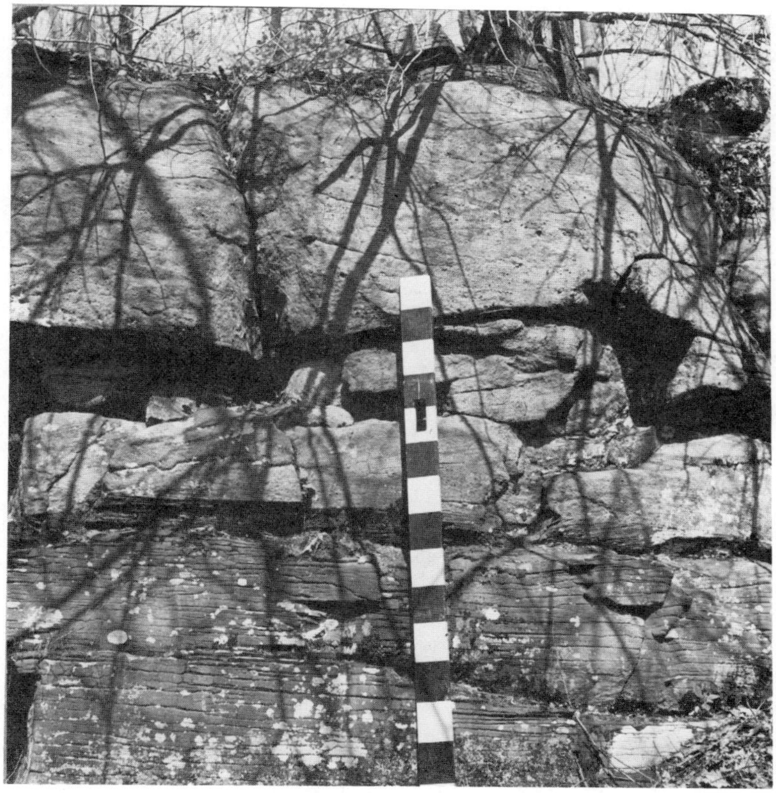

Figure 31. Conglomeratic sandstone in the upper part of the Delaware River Member about 0.9 mile east of the north end of Walker Lake at 41°25′25″N/74°53′38″W. The conglomeratic zone is the upper half of the exposure. The scale is divided in 10-cm intervals.

0.7 mm and comprises mainly highly strained quartz, polycrystalline quartz, shale, siltstone, and metamorphic rock fragments. The weathered feldspar content is slightly higher in the conglomeratic sandstone than in the sandstone. Higher percentages of microcrystalline silica and authigenic silica overgrowths occur in the conglomeratic sandstone and are a result of greater solution activity in the coarser grained rocks. Heavy minerals are almost totally absent, and chloritization of illite is either absent or minor. The one sample (not in Table 2) of conglomeratic sandstone from western Pike County (East Mountain) differs from the samples from eastern Pike County mainly in its low feldspar content.

The intraformational conglomerates are strikingly different from the other rocks because of their low quartz content and their large amount of carbonate clasts and calcite cement. The carbonate clasts are rounded, micritic masses generally several times larger than the other framework components. Sparry-calcite cement dominates, and micrite cement is absent in more than half of the samples. Considerable alteration to calcite has occurred in the feldspars. In most samples there is a minimal number of point contacts between framework components, which are essentially floating in sparry calcite. No heavy minerals occur, and chloritization of illite is minimal to absent.

THICKNESS: The maximum thickness of the Delaware River Member in eastern Pike County, where it is overlain by the Lackawaxen Member,

Figure 32. Conglomerate typical of the conglomeratic zone in the upper part of the Delaware River Member. The specimen is from the exposure shown in Figure 31. The scale is divided in inches and centimeters.

is calculated from cross section A–A′ (Plate 1) to be 2,600 feet. In the western part of the county, where the Lackawaxen Member is absent, the maximum thickness should be about 2,800 feet.

CONTACTS: The base of the Delaware River Member is placed at the base of the first red-bed unit in the Catskill Formation sequence. The first red bed is not presumed to occur in exactly the same stratigraphic position across the county, but mapping indicates a reasonable persistence at about the same level. The upper contact is placed at the bottom of the basal conglomeratic zone of the overlying Lackawaxen Member where it is present and at the bottom of the basal conglomeratic zone of the undivided Poplar Gap and Packerton Members where the Lackawaxen is absent.

EXPOSURES: Good exposures of the Delaware River Member are abundant in the numerous flagstone quarries in the northeastern part of the county (see quarry symbols on Plate 1). Specific locations of several good exposures are given in the captions of Figures 22 to 31. Excellent exposures occur at Shohola Falls and also form the near-vertical walls of the narrow gorge immediately below the falls. Good exposures occur below the Lackawaxen Member in numerous places along the south bank of the Lackawaxen River, particularly at Glen Eyre, Rowland, and Lackawaxen. A good quarry exposure occurs on the west side of Pecks Pond at 41°17′21″N/ 75°05′26″W. The best exposures in the western part of the county are on the steep slope of the Pocono Plateau escarpment in the area of East Mountain.

ORIGIN: The Delaware River Member was deposited by shallow braided streams that flowed to the north and northwest from two distinct centers of sediment dispersal and that headed many miles to the southeast in the Acadian mountains. The separate sediment-dispersal systems have been given the local names Egypt Mills and Port Jervis. The Port Jervis depositional system contained abundant amounts of sand; most fine-grained sediments were apparently transported farther onto the alluvial plain and were deposited elsewhere. The braided streams were subjected to repeated episodes of high-flow-regime transport during which time the planar-bedded sandstone was deposited. The feldspar content of the sediments deposited by this system distinguishes them from those deposited by the Egypt Mills system.

The Egypt Mills depositional system had a slightly different source terrain, and its sediments are basically feldspar deficient. The gradient of the streams and the amount of sand available were apparently different also, with the result that more fine-grained sediments were deposited in flood basins adjacent to the main braided-stream network. Either restrictions on the sand supply or a lower flood-flow velocity resulted in the deposition of fewer planar-bedded sandstones. There is no readily identifiable boundary between the rocks that originated from each of the different sediment-dispersal systems, and there was probably a considerable amount of lateral intermixing during the course of their deposition.

The conglomeratic zones near the top of the member presumably reflect short pulses of tec-

Table 2. *Microscopic Composition of the Delaware River Member (86 Samples)*

(Values represent percentage of total composition)

| Component | Sandstone (east) (61 samples) | Sandstone (west) (14 samples) | Conglomeratic sandstone (east) (4 samples) | Intra-formational conglomerate (7 samples) |
|---|---|---|---|---|
| Framework | 68 | 78 | 76 | 62 |
| Cement | 29 | 17 | 24 | 37 |
| Matrix | 3 | 3 | Tr[1] | 1 |
| Framework | | | | |
| Strained quartz | 50 | 57 | 41 | 22 |
| Feldspar | 4 | 1 | 5 | 2 |
| Polycrystalline quartz | 2 | 2 | 14 | 2 |
| Stretched metaquartz | Tr | Tr | Tr | 1 |
| Shale clasts | 6 | 11 | 5 | 3 |
| Siltstone clasts | Tr | Tr | 3 | 1 |
| Metamorphic rock fragments | 2 | 4 | 5 | 2 |
| Coarse micas | 3 | 2 | 2 | Tr |
| Carbonate clasts | 0 | 0 | 0 | 29 |
| Miscellaneous | 1 | 1 | 1 | Tr |
| Cement | | | | |
| Illite-chlorite-chert composite | 19 | 13 | 12 | 4 |
| Microcrystalline silica | 4 | 2 | 6 | 2 |
| Authigenic silica overgrowths | 5 | 2 | 6 | 2 |
| Calcite | 1 | Tr | 0 | 29 |
| Matrix | | | | |
| Kaolinite, illite, chlorite, and limonite | 3 | 3 | Tr | 1 |

[1]Tr, trace.

tonism in the source areas, which resulted in brief flushings of coarser sediment onto the alluvial plain. The tectonic events did not persist long enough to allow deposition of a widespread gravel blanket comparable to those of the overlying Lackawaxen or Poplar Gap and Packerton Members.

MISCELLANEOUS: The planar-bedded sandstone in the northeastern part of the county was once quarried extensively for flagstone (bluestone), and abundant quantities of this material still remain. The various aspects of flagstone are discussed in the section "Mineral Resources." Some of the sandstone has been quarried for use as crushed rock, but none is being so utilized today. The red siltstone and claystone are generally not thick enough to warrant development as a source of random fill, but some local use of this rock has occurred.

## Lackawaxen Member

INTRODUCTION: The term Lackawaxen conglomerate was applied to "a very massive pebbly bed 50′–60′ thick, which occurs near the top of this series [Delaware river flags] in the vicinity of Lackawaxen" by I. C. White in 1882 (p. 100) (see Figure 3). White recognized the presence of this conglomerate at numerous localities in the northern third of Pike County and apparently considered it to be only one stratigraphic horizon. He later ascribed regional significance to this conglomerate and suggested that it occurred in Huntingdon County, Pennsylvania (White, 1885, p. 90–91). Although White seemed positive about the correlation, Willard (Willard and others, 1939, p. 299) dismissed the correlation outright. Willard apparently attached no significance to the Lackawaxen conglomerate even in Pike County and did not include the name in his stratigraphic scheme (p. 292–294). He placed the conglomerate interval in his Shohola Formation.

Fletcher and Woodrow (1970, p. 20–21) recognized the Lackawaxen conglomerate as a part of a mappable rock unit overlying their Shohola Formation. They named the unit the Lackawaxen Formation and defined the base of the formation as the base of the lowest mappable conglomerate (although they wrote that the basal contact is gradational through an interval at least 100 feet thick). They did not define a top for the unit, but estimated its thickness as much greater than 600 feet. Fletcher and Woodrow suggested that the formation correlates with the lower part of the Slide Mountain Formation in eastern New York (Fletcher, 1963; Rickard, 1975) and the upper part of the Walton Formation in east-central New York (Fletcher, 1963). Regional reconnaissance by Sevon and Berg supports the former correlation.

Geologic mapping for this report shows that the Lackawaxen Formation is not a valid unit as defined and mapped by Fletcher and Woodrow. The reason for this is that there are several conglomeratic zones in northwestern Pike County in the unit they defined as the Lackawaxen Formation. Neither Fletcher and Woodrow nor White were apparently aware of this complication. The conglomeratic zone that Fletcher and Woodrow used to define the base of the formation is only locally developed (it cannot be traced west of Shohola Creek north of Shohola Falls) and in this report is included in the Delaware River Member. A regionally mappable conglomeratic zone occurs at a higher position and is the same one originally named by White. A higher conglomeratic zone also occurs, but is not as persistent. In this report the Lackawaxen is redefined as a member of the Catskill Formation, and its base is placed at the base of the mappable conglomeratic zone. The zone is well exposed in the area around Lackawaxen and can be traced as far west as Mud Pond Mountain to the south of Promised Land Lake. In the western part of Pike County the top of the redefined Lackawaxen Member is placed at the base of the overlying Poplar Gap and Packerton Members. Arbitrary boundaries for the member are drawn west of Mud Pond Mountain and at a spur above Lake Wallenpaupack just east of Ironwood Park near the west end of the lake (see Plate 1). Rocks considered characteristic of the Lackawaxen have not been located to the west of these boundaries. The Lackawaxen Member is the uppermost stratigraphic unit in the remainder of the county.

The Lackawaxen Member is characterized by the occurrence of conglomeratic zones and a moderate quantity of detrital feldspar. The feldspar content is higher than it is in stratigraphically equivalent units to the southwest and thus aids in the delineation of the Lackawaxen as a distinct member.

FIELD OCCURRENCE: The Lackawaxen Member crops out in the northern part of Pike County, primarily in Lackawaxen, Blooming Grove, and Shohola Townships (see Plate 1). Because the unit is dominated by sandstone, ledges are a characteristic landform. The conglomeratic zones and adjacent sandstones are particularly resistant, forming steep-faced ledges many feet high along parts of the Lackawaxen River and prominent ledges in areas removed from major drainage channels. Ledge faces generally parallel fracture systems, and ledge development has been enhanced by glacial erosion. The ground surface downslope from a ledge commonly has moderate to large quantities of broken rock derived from the ledge. In some cases very large blocks have been detached by glacial action and have moved several feet from the ledge (Figure 33).

DESCRIPTION: Most of the Lackawaxen Member is gray, crossbedded, feldspathic sandstone. Some red siltstone and claystone occur, particu-

Figure 33. Detached blocks of sandstone from the Lackawaxen Member. The exposure is along the east side of a private road between U.S. Route 6 and Lake Giles about 0.6 mile due south of U.S. Route 6 at 41°24′31″N/ 75°05′48″W.

larly in the northwest half of the area of its occurrence. These siltstones and claystones probably make up less than 10 percent of the member, are rarely seen in outcrop, and do not appear to be different from those described for the underlying Delaware River Member. Two thin conglomeratic zones occur, one at the base and the other near the top of the unit.

The sandstone ranges from very fine to coarse grained, and variations in grain size between adjacent beds are common. Fine- to medium-grained sandstone is most common and in many places occurs as sequences of fine-grained, planar-bedded sandstone overlain by medium-grained, crossbedded sandstone.

The planar-bedded sandstone has been quarried for flagstone at numerous localities and near the surface has a tendency to part parallel to bedding at a spacing of about 1 to 2 inches (Figure 34). Fresh rock generally has a thicker parting tendency. Individual laminae are 1 to 2 mm thick. Parting-step lineations are common on bedding surfaces, and carbonaceous fragments of plant material occur in local concentrations or randomly scattered on some bedding surfaces. Small pyrite cubes locally occur in association with the carbonaceous material, and limonite staining commonly accompanies weathering of the fragments. Except for the usually visible laminae and the tendency for parting along some laminae, the planar-bedded sandstone is quite uniform in appearance and may be up to 6 feet thick in an uninterrupted interval. The upper surface of the sandstone is sharp and in some places irregular, having relief up to several inches.

Overlying the planar-bedded sandstone is coarser grained, crossbedded sandstone. The crossbedded sandstone is generally medium grained, but some coarse- to very coarse grained beds occur. Scattered white quartz pebbles up to 1.25 cm in diameter and dark-gray platy shale clasts up to 8 cm in diameter occur near the base of some crossbeds. Rarely, a concentration of these clasts occurs in association with carbonaceous debris as an intraformational conglomerate. The intraformational conglomerates are seldom more than 8 to 12 inches thick and 6 feet long. The conglomerates have calcite cement and develop recessed areas in outcrop because they weather more rapidly than the adjacent sand-

Figure 34. Planar-bedded sandstone of the Lackawaxen Member exposed along the east side of Blooming Grove Creek, about 1.4 miles southwest of Glen Eyre at 41°27′00″N/ 75°05′34″W. The hammer (center, on top of the log) is 11 inches long.

stone. Individual crossbedded units are 1.5 to 3 feet thick. Commonly, the crossbedding immediately above the planar-bedded sandstone is low angle, has a tangential base, and is relatively uniform in appearance. Above this the crossbedding is almost exclusively of the trough type (Figure 35); the troughs are generally less than 6 feet wide and 1.5 feet thick. Boundaries between crossbedded units are sharp. On many natural exposures, the crossbeds are well etched by weathering, revealing individual beds 0.5 to 2 inches thick and laminae 1 to 5 mm thick. Laminae are difficult to detect in the coarser grained sandstone.

The breakup of sandstone in the Lackawaxen Member is controlled primarily by joints and bedding sets and secondarily by bedding partings. Initially the sandstone breaks into blocks of various sizes. These blocks gradually break up into platy pieces along weakly cemented bedding surfaces. Granular disintegration is minimal.

The base of the Lackawaxen Member is defined by an areally persistent and mappable conglomeratic zone (Figure 36). The zone is up to 20 feet thick and consists of mainly coarse- to very coarse grained, feldspathic, conglomeratic sandstone containing some conglomerate beds 3 to 5 feet thick. The sandstone beds are 1.5 to 3 feet thick, but in many places are so well cemented that several sets of beds give the appearance of a single massive bed. Bed sets that are well etched by weathering have sharp boundaries between beds, but generally show little evidence of significant scour at the base of any bed. All of the beds have trough crossbedding as the primary sedimentary structure, but cementation obscures the crossbedding in some locations. Crossbed laminae are generally 1 to 2 mm thick, and some adjacent laminae have moderate differences in grain size.

Figure 35. Trough-crossbedded sandstone in the Lackawaxen Member exposed along the west side of Blooming Grove Creek about 1.4 miles southwest of Glen Eyre at 41°27′00″N/75°05′34″W. The arrows are aligned along the centers of the troughs and point in the direction of the depositional current flow (predominantly N55°W).

Generally, less than 15 percent of the conglomerate beds are clasts larger than 2 mm in diameter, but locally the percentage is as much as 75. The clasts are dominantly rounded to subangular, white quartz averaging 0.5 to 1.25 cm in diameter and having a maximum observed size of 2.5 cm. Some sandstone, siltstone, shale, and quartzite clasts occur, but they are a minor component of material larger than 2 mm in diameter. Some beds consist dominantly of sand-sized grains and a few scattered quartz pebbles. Feldspar grains found in the conglomerate beds are sand sized, but are generally easily recognizable on freshly broken surfaces.

Breakup of the conglomeratic beds is controlled mainly by joints and partings between bed sets and results in the development of blocks of various size. Long-continued disintegration results in fragmentation along crossbed partings, and platy pieces are produced.

A second conglomeratic zone occurs about 300 feet above the base of the Lackawaxen Member, but probably varies somewhat in stratigraphic position from place to place. This zone consists of apparently laterally discontinuous beds of dominantly coarse- to very coarse grained, feldspathic sandstone containing scattered quartz pebbles and one or more beds of conglomerate containing up to 50 percent pebbles. The sandstone beds are generally 2 to 4 feet thick and have sharp bed-set boundaries. Trough crossbedding

Figure 36. Basal conglomeratic zone of the Lackawaxen Member exposed near the north end of the spur between the Lackawaxen and Delaware Rivers in Lackawaxen at 41°28′45″N/74°59′44″W.

occurs throughout these beds (Figure 37), but is in some places obscured by cementation, which causes several bed sets to have a massive, structureless appearance. Scattered quartz pebbles are up to 4 cm in diameter, but generally average less than 0.6 cm.

The conglomerate beds of the upper conglomeratic zone are generally less than 3 feet thick and consist mainly of rounded to subangular quartz pebbles up to 4 cm in diameter, which have an average size of about 0.6 cm (Figure 38). Some pebbles of siltstone and quartzite occur, as do a few rounded, flat, dark shale clasts up to 5 cm in long dimension. The larger pebbles are commonly concentrated along laminae near the base of a bed. The beds are trough crossbedded, but the crossbedding is obscure in some places. The conglomerate beds occur near the top of the zone.

This zone, which is up to 10 feet thick, occurs as a ledge-forming unit near the tops of several small hills in the northwestern third of the county (see Plate 1). The isolated occurrences of the zone and the absence of similar rock in areas of comparable stratigraphic position indicate that this unit was not deposited as a widespread sheet in the manner of the basal conglomeratic zone, but instead was restricted to a few channels.

Breakup of this zone is similar to that of the basal conglomeratic zone.

MICROSCOPIC PETROLOGY: Point-count analyses of 56 thin sections of sandstone and 18 thin sections of conglomeratic sandstone from the Lackawaxen Member are summarized in Table 3.

The dominant framework component of the sandstone is detrital quartz, which is moderately

Figure 37. Trough-crossbedded conglomeratic sandstone in the upper conglomeratic zone of the Lackawaxen Member exposed about 0.7 mile east of High Falls at 41°24′24″N/75°05′28″W. The base of the conglomeratic bed is above the hammer head. The hammer (left center) is 11 inches long.

to highly strained. These grains are subangular to subrounded, and the majority of them have probably had their shapes modified by secondary marginal solution or precipitation. Minute traces of grain coatings occasionally occur within authigenic silica overgrowths and define the original shape of moderately well rounded grains. In some of the thin sections of planar-bedded sandstone, there is a pronounced orientation of the long axes of quartz grains parallel to bedding. Micas and elongate rock fragments are similarly aligned. Various rock fragments and detrital micas (mainly muscovite, some biotite, and rare chlorite) constitute much of the rest of the sandstone framework. The rock fragments are commonly crushed and distorted and have indefinite outlines, which makes identification difficult. Feldspar occurs as both orthoclase and plagioclase in variable proportions, but frequently there are larger amounts of orthoclase. All of the feldspar is weathered, and some is almost entirely altered to clay or replaced by calcite. Heavy minerals are not abundant, except in some rocks where they are concentrated along single laminae. Rounded zircons are most common, but garnet, apatite, magnetite, and ilmenite occur. The binding material is a homogeneous mass of illite and microcrystalline silica, which totally fills all of the framework interstices and is often difficult to distinguish from weathered feldspar and crushed rock fragments. Some calcite cement occurs, but it is a minor constituent. The illite fraction has been chloritized to some extent in all of the sandstone thin sections examined. Chloritization is well advanced in some samples, but no areal trends are apparent.

The conglomeratic zones contain substantially more framework components than the sand-

Figure 38. Conglomerate bed in the upper conglomeratic zone of the Lackawaxen Member exposed about 0.25 mile northwest of High Falls at 41°06′22″N/75°01′54″W. The scale is divided in inches and centimeters.

stone. The larger quantities of feldspar and rock fragments present may reflect a greater survival potential for these grains in the larger sizes. This is probably particularly true for the polycrystalline quartz and stretched metaquartz grains, which reach a critical size and degree of weathering at which they disintegrate into component pieces. Feldspar grains are angular and weathered, but the weathering product is generally masked by limonite staining and calcite is absent as a replacement component. The proportions of orthoclase and plagioclase are about the same as in the sandstone. Metamorphic rock fragments appear to be mainly gneisses. The quartz grains are all strained, and many have etched margins. No heavy minerals were noted in these thin sections. The degree of chloritization is variable, but it is less than in the sandstone and almost absent in some of the thin sections.

THICKNESS: The maximum thickness of the Lackawaxen Member in Pike County is estimated to be between 400 and 500 feet in the northern third of the county, where the member is not overlain by a higher stratigraphic unit, and about 200 feet in the western part of the county, where it intertongues with the undivided Poplar Gap and Packerton Members. The interval between the base and the upper conglomeratic zone is generally about 300 feet, but is probably several feet more or less in places. A few hills have up to 150 feet of sandstone above the upper conglomeratic zone.

CONTACTS: The base of the Lackawaxen Member is placed at the base of an areally mappable conglomeratic zone, the lower conglomeratic zone of this report. This conglomeratic zone is very persistent except in the western part of the county, where the member interfingers with the overlying undivided Poplar Gap and Packerton Members. A boundary for the member is placed west of Mud Pond Mountain be-

Table 3. *Microscopic Composition of the Lackawaxen Member (74 Samples)*

(Values represent percentage of total composition)

| Component | Sandstone (56 samples) | Conglomeratic sandstone (18 samples) |
|---|---|---|
| Framework | 65 | 75 |
| Cement | 33 | 25 |
| Matrix | 2 | 1 |
| Framework | | |
| Strained quartz | 48 | 45 |
| Feldspar | 5 | 15 |
| Polycrystalline quartz | 1 | 5 |
| Stretched metaquartz | Tr[1] | 2 |
| Shale clasts | 4 | 1 |
| Siltstone clasts | Tr | 2 |
| Metamorphic rock fragments | 1 | 3 |
| Coarse micas | 4 | 1 |
| Miscellaneous | 2 | 1 |
| Cement | | |
| Illite-chlorite-chert composite | 19 | 11 |
| Microcrystalline silica | 6 | 9 |
| Authigenic silica overgrowths | 6 | 5 |
| Calcite | 2 | 0 |
| Matrix | | |
| Kaolinite, illite, chlorite, and limonite | 2 | 1 |

[1]Tr, trace.

cause the conglomeratic zone has not been located to the southwest on the slopes of East Mountain.

The upper contact in the western part of the county is placed at the base of the basal conglomeratic zone of the overlying undivided Poplar Gap and Packerton Members. Throughout the remaining area of its occurrence in Pike County, the member is not overlain by a higher rock-stratigraphic unit and has no upper contact. Insufficient reconnaissance has been done north and northwest of Pike County to ascertain the extent and mappability of the member outside the county.

An arbitrary boundary defines the western limit of the Lackawaxen Member near the southwestern end of Lake Wallenpaupack. Rocks to the west of this boundary are feldspar deficient, and rocks to the east are feldspathic. However, no other change in the character of the rocks is readily discernible.

EXPOSURES: Exposures of the Lackawaxen Member are moderately abundant throughout the area. The rocks occur as ledges, which have developed along fractures and have been scraped by glaciation. Many of the ledges are broken and disrupted as the result of disintegration along fractures, but some, particularly the larger ledges, are unbroken. Most of the ledges have been etched by weathering so that bedding and cross-bedding are well displayed.

The lower conglomeratic zone crops out discontinuously along moderate to steep slopes in the northwestern half of the county. On moderate slopes, the ledges are generally only 3 to 7 feet high, and in some places, the presence of conglomerate is indicated by float blocks below the ledge. On steep slopes, such as along the south side of the Lackawaxen River, the ledges are commonly more than 10 feet high, and most of the zone is exposed. Good exposures of this unit occur along the south side of the Lackawaxen River at Rowland (41°28′17″N/75°02′21″W) and at Lackawaxen (41°28′45″N/74°59′44″W and 41°28′32″N/74°59′47″W); east of Lackawaxen near Panther Creek (41°27′58″N/74°57′24″W); along the eastbound lane of Interstate Route 84 near Blooming Grove (41°22′01″N/75°08′21″W); along the power line on the south side of High Knob (41°17′48″N/75°07′24″W); and on the south side of Mud Pond Mountain (41°16′45″N/75°11′03″W).

The upper conglomeratic zone crops out as ledges at or near the tops of small hills in the northwestern part of the county. Good exposures of this unit occur at the north end of White Deer Lake (41°24′39″N/75°07′57″W); near the top of Big Hill west of White Deer Lake (41°24′42″N/75°09′54″W); north of Masthope Creek (41°34′01″N/75°02′53″W); west of Paupack (41°24′01″N/75°13′07″W); and southwest of Germantown near High Falls (41°24′24″N/75°05′28″W).

Good exposures of Lackawaxen sandstone occur throughout the area as ledges, particularly on the slopes south of Masthope Creek and the Lackawaxen River. Good exposures also occur in the many flagstone quarries opened in this unit. Readily accessible exposures occur in a large quarry adjacent to U.S. Route 6 southwest of Taylortown (41°24′19″N/75°01′24″W); at High Falls (41°24′12″N/75°06′15″W); along the bed of Blooming Grove Creek; along the main road past Corilla and Forest Lakes; along the power line on the south side of High Knob; and on the slopes above the upper conglomerate at the north end of White Deer Lake.

The siltstone is rarely exposed naturally and is best seen in borrow pits. Exposures occur along the road adjacent to Masthope Creek at the base of Point Peter (41°33′19″N/75°04′06″W); in a borrow pit on the south side of the road at Paupack (41°23′38″N/75°11′43″W); in a borrow pit on the east side of Pa. Route 390 north of Tafton (41°24′33″N/75°10′55″W); along and at the end of the road to Lake Earnest; and along Blooming Grove Creek just north of High Falls.

ORIGIN: The Lackawaxen Member was deposited primarily by shallow, braided streams flowing to the northwest from the Acadian mountains many miles to the southeast. This sediment-dispersal system is given the local name Port Jervis. The highlands probably contained gneisses, which contributed abundant amounts of feldspar to the sediment load in the streams. The widespread occurrence of the lower conglomeratic zone presumably corresponds to an extended period of coarse-sediment supply, during which time the braided streams deposited a thin gravel unit over the entire alluvial plain. In contrast, the apparently restricted nature of distribution of the upper conglomeratic zone suggests a shorter period of coarse-sediment supply, during which time gravel was deposited locally, but not as a widespread unit.

The planar-bedded sandstone from which the flagstones are quarried presumably represents deposition of abundant amounts of sand during periods of high-velocity flow.

Flood-basin areas existed locally, and silts and clays were deposited in these areas periodically by floods. During subsequent periods of quiescence, vegetation and a variety of small terrestrial invertebrates were established and the fine-grained sediments were oxidized (resulting in a red color). These areas probably resembled small, shallow lakes immediately following a flood, but were dry within a short time after the flood.

MISCELLANEOUS: The Lackawaxen Member has been a source of flagstone, and this use is discussed in the "Mineral Resources" section. The red siltstone and claystone have been utilized as random fill, primarily for low-use secondary roads, and that use is also discussed in detail in the section "Mineral Resources."

## Poplar Gap and Packerton Members, Undivided

INTRODUCTION: The Poplar Gap and Packerton Members of the Catskill Formation are mapped as separate units southwest of Pike County (e.g., Sevon, 1975a), but are considered inseparable for mapping purposes in Pike County. The Packerton Member was originally defined along the Lehigh River about 37 miles southwest of Pike County (Epstein and others, 1974). This resistant unit forms the topographic edge of the Pocono Plateau from the Lehigh River to Pike County. The Poplar Gap Member was defined by Berg (1975) and represents a sequence of rocks above the Packerton Member and below the Duncannon Member of the Catskill Formation. On the Lehigh River this sequence is composed of (from youngest to oldest) the Clarks Ferry, Berry Run, and Sawmill Run Members of the Catskill Formation, which become undifferentiable northeast of Carbon County. The Poplar Gap Member was defined to encompass this sequence of rocks. As subdivision of the Catskill Formation continues farther north in eastern Pennsylvania, this sequence should probably be redefined and given a new name, but that has not been done in Pike County because only the lower part of the sequence is exposed.

The Poplar Gap and Packerton Members are similar in appearance to the Lackawaxen Member, but are compositionally dissimilar because of their lack of feldspar, which is characteristic of the Lackawaxen.

FIELD OCCURRENCE: The Poplar Gap and Packerton Members occur in the western part of Pike County south of Lake Wallenpaupack, west of High Knob, and north of the Pocono Plateau escarpment in Green and Blooming Grove Townships (see Plate 1). The members form the resistant front edge of the Pocono Plateau escarpment, which is the dominant topographic feature southwest of High Knob. The unit is readily traceable north of High Knob, but is not associated with prominent topography.

DESCRIPTION: Most of the undivided Poplar Gap and Packerton Members is gray, crossbedded sandstone. Some red siltstone and claystone occur, but they are not well enough exposed to evaluate how much of the unit they compose. The base of the members is marked by a resistant conglomeratic zone, and other conglomeratic beds occur higher in the unit. The unit is readily distinguishable from similar appearing rocks of the Lackawaxen Member because the Poplar Gap and Packerton Members contain almost no feldspar.

The sandstone ranges from very fine to very coarse grained, and variations in grain size between adjacent beds are common. Medium- to coarse-grained sandstone is most common, and very fine grained sandstone is rare. The sandstone is generally gray to olive gray where fresh and weathers to very light gray or almost white. Trough crossbedding pervades the sandstone (Figure 39); the troughs are up to 10 feet or more wide and generally less than 3 feet deep. Individual crossbed units range from 1.5 to 3 feet thick and are separated from adjacent beds by sharp boundaries. Although the crossbedding generally has laminations at a scale of a few millimeters, most weathered outcrops display crossbed partings ranging in width from 1 to 10 cm.

Scattered quartz pebbles up to 2.5 cm in diameter occur in most of the sandstone, and thin concentrations of pebbles occur along laminae near the base of some of the crossbeds. Pebbles are most common in sandstone near conglomeratic zones.

Natural disintegration of the sandstone is first into blocks of various sizes controlled by joint and bedding partings and ultimately into slabby or platy pieces, the sizes of which are controlled by crossbed partings.

Local concentrations of calcareous intraformational conglomerate occur within the sandstone. These concentrations are generally lens-

Figure 39. Trough-crossbedded sandstone in the Poplar Gap and Packerton Members exposed about 0.45 mile northeast of the crest of Mud Pond Mountain at 41°17′08″N/75°10′14″W. The hammer is 11 inches long.

oidal to irregular in plan, generally less than 1.5 feet thick and 3 to 6 feet long, and are readily distinguished by either the brownish color on weathered surfaces of recent exposures or by the blackish color, recessed weathering, and friable nature of older exposures. The intraformational conglomerates comprise numerous flat, green or gray shale clasts up to several centimeters long, some quartz pebbles, some carbonaceous plant fragments, and numerous round or rod-shaped calcareous nodules, all enclosed in a sand matrix and cemented with calcite. These concentrations appear to be localized in the bases of trough-crossbed channels and apparently have no predictable lithologic or stratigraphic associations. The bases of the intraformational conglomerates are sharp, and the lateral and upper boundaries are transitional into adjacent sandstone. Good exposures of this material occur in roadcuts along the eastbound lane of Interstate Route 84 at 41°21′38″N/75°12′36″W and 41°21′49″N/75°09′20″W.

The base of the Poplar Gap and Packerton Members is marked by an areally persistent conglomeratic sandstone generally 3 to 6 feet thick. This zone is very well cemented and forms the erosion-resistant edge of the Pocono Plateau southwest of High Knob. The conglomeratic sandstone contains mainly rounded to subrounded, white quartz pebbles in concentrations ranging from 10 to 75 percent. Various rock fragments and a few red quartz pebbles also occur. The red quartz seems to be characteristic of this unit. The average size of the pebbles is 0.5 to 1 cm, and the maximum size is generally less than 4 cm. The pebbles vary from thin concentrations along the bases of crossbeds to relatively uniform distributions throughout a crossbed set (Figure 40). The conglomeratic beds are probably all trough crossbedded (Figure 41), but cementation of the beds is so strong in most places that weathering is insufficiently advanced to show the crossbeds, and the beds have a structureless appearance. Where crossbedding is observable, the crossbed dip appears to be steeper than that of the crossbeds in the sandstone. In the few places where the size of crossbed troughs were observed, they were greater than 10 feet in width.

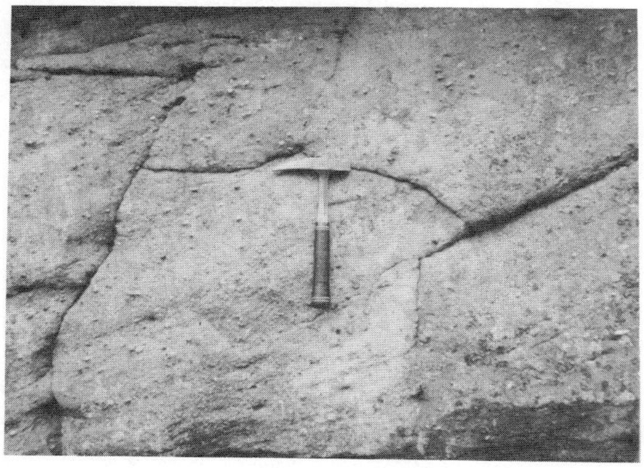

Figure 40. Part of the basal conglomeratic sandstone of the Poplar Gap and Packerton Members exposed about 0.75 mile southwest of Blooming Grove at 41°21′39″N/75°09′38″W. The hammer is 11 inches long.

Figure 41. Trough-crossbedded, basal conglomeratic sandstone in the Poplar Gap and Packerton Members exposed about 0.5 mile west of Lake Earnest at 41°17′19″N/75°10′10″W. An 11-inch-long hammer is located in the middle of the vertical rock face to the right of center.

Conglomeratic zones occur above the basal conglomeratic zone and are basically similar to the basal unit except that the upper conglomeratic beds are thinner, generally less than 3 feet thick, and seldom contain more than 30 percent pebbles. The average pebble size is about 0.5 to 1 cm, and the maximum size is 5 to 8 cm. Rounded white quartz pebbles dominate the coarse fraction, but red quartz pebbles are commonly noticeable. There does not appear to be any stratigraphic regularity in the occurrence of the higher conglomeratic zones.

All of the conglomeratic sandstones are very well cemented and resistant to natural disintegration. Initial breakup is controlled by joint and bed-set partings and produces moderate to large blocks. Subsequent disintegration is controlled by crossbed partings and produces slabby or platy pieces. The rate of weathering of these rocks by granular disintegration is slow.

All of the conglomeratic sandstones are medium gray where fresh and have a slight hint of green. The more conglomeratic zones tend to appear almost white where weathered because of the concentration of white quartz pebbles, but weathering of iron near the surface gives the sandstone a subtle grayish-red coloration.

Grayish-red siltstone and claystone occur in the Poplar Gap and Packerton Members, but are not well exposed in Pike County except in artificial cuts along Interstate Route 84 and in two borrow pits near Lake Russell. At the base of these fine-grained sequences, there is a thin transitional interval through which gray, fine-grained sandstone grades upward into very fine grained, grayish-red sandstone, which grades upward into the grayish-red siltstone and claystone. The siltstone and claystone occur in cyclic alternations. The siltstone is fissile to subfissile except where fissility has been destroyed by burrowing. The claystone is almost everywhere totally homogenized by burrowing and shows no bedding structure.

Subhorizontal to vertical tubelike structures filled with homogenized sediment are common in the siltstone and claystone that display bedding laminations. These tubes, which represent burrows formed by organisms of unknown affinity, range from 2 mm to about 3 cm in diameter. The frequency of occurrence of these tubes commonly increases upward from a few scattered tubes into beds which have been so totally homogenized by burrowing that individual tubes are not recognizable and bedding is obliterated. Larger vertical tubelike structures about 5 to 10 cm in diameter occur, but are not common. These structures are probably the aestivation burrows of dipnoan lungfish. Small, in some places angular, fragmentlike, light-colored clay films are common and represent remnant traces of oxidized plant material, principally roots. In rare cases, a branching pattern is preserved. Faint impressions of leaves from the plant *Archaeopteris* occur, but are rarely found because of the difficulty of detection and the ease with which the fine-grained rocks disintegrate.

The siltstone and claystone disintegrate readily in natural and artificial exposures. Where original bedding is present the rock breaks up into thin, platy pieces. More commonly, where bedding is absent because of burrowing, the rock breaks up into small, hackly and chippy fragments a few centimeters or less in long dimension.

MICROSCOPIC PETROLOGY: Point-count analyses of 20 thin sections of sandstone from the Poplar Gap and Packerton Members are summarized in Table 4. The average grain size for all of the thin sections is 0.6 mm.

The dominant framework component is detrital quartz, which is moderately to highly strained. The quartz grains are subangular to subrounded, but many of the grains have had their shapes modified by secondary marginal solution or silica overgrowth. Rounded grains of polycrystalline quartz make up the next most abundant detrital component, and elongate, rounded grains of shale, siltstone, and metamorphic rock fragments (gneiss?) make up most of the rest of the framework. Coarse-grained detrital micas, both muscovite and biotite, occur, but are not abundant. Feldspar and heavy minerals are conspicuous by their general absence.

Illite and microcrystalline quartz constitute most of the cementing material in the sandstone, although calcite is important in some rocks. Many quartz grains are bound to adjacent grains by silica overgrowths, and in some cases solution-modified boundaries are interlocked. Chloritization of the illite is variable; it is seldom pervasive, and in some places it is absent.

Table 4. *Microscopic Composition of the Poplar Gap and Packerton Members, Undivided (20 Samples)*

(Values represent percentage of total composition)

| Component | Sandstone (20 samples) |
|---|---|
| Framework | 77 |
| Cement | 22 |
| Matrix | Tr[1] |
| Framework | |
| Strained quartz | 55 |
| Feldspar | 1 |
| Polycrystalline quartz | 9 |
| Stretched metaquartz | 1 |
| Shale clasts | 4 |
| Siltstone clasts | 3 |
| Metamorphic rock fragments | 2 |
| Coarse micas | 1 |
| Miscellaneous | 1 |
| Cement | |
| Illite-chlorite-chert composite | 9 |
| Microcrystalline silica | 6 |
| Authigenic silica overgrowths | 4 |
| Calcite | 3 |
| Matrix | |
| Kaolinite, illite, chlorite, and limonite | Tr |

[1]Tr, trace.

THICKNESS: The maximum thickness of the Poplar Gap and Packerton Members in Pike County is estimated to be 500 feet in the vicinity of Carlton Hill and Lake Russell. Throughout much of the area, particularly in the vicinity of Promised Land State Park, the thickness is probably less than half of the maximum.

CONTACTS: The base of the Poplar Gap and Packerton Members is placed at the base of an areally persistent conglomeratic zone, which forms the erosion-resistant edge of the Pocono Plateau southwest of High Knob and is readily traceable to the north and west of High Knob. The actual contact is not placed on a specific bed, but is placed at the ledge containing the basal conglomeratic zone. In some places this zone is underlain by grayish-red siltstone and claystone, which allows easy placement of the contact. Where the underlying bed is sandstone, placement of the contact is arbitrary and depends largely on lateral continuity with areas having more certain boundaries.

The upper contact of the unit does not occur in Pike County.

EXPOSURES: Exposure of the Poplar Gap and Packerton Members is plentiful along the Pocono Plateau escarpment in the southern part of the area of occurrence; moderately abundant east, north, and in the vicinity of Promised Land State Park; and least abundant in the western part of the county. The exposures occur as ledges, the orientations of which are controlled by jointing and the sizes of which are largely proportional to the amount of glacial erosion and deposition that has occurred.

The most accessible exposures occur in roadcuts along Interstate Route 84. A good roadcut along the eastbound lane northwest of Bruce Lake at 41°21′30″N/75°10′51″W displays a variety of lithologies. A good exposure of conglomeratic sandstone occurs as a ledge on a small hill immediately north of the westbound lane at 41°21′42″N/75°09′55″W (northeast of Bruce Lake). Other moderately good exposures of the unit occur along both lanes of the highway farther west.

The best exposures of the basal conglomeratic sandstone occur along the Pocono Plateau escarpment, but these are on private property and are not readily accessible except on foot. These exposures include those in the vicinity of Mud Pond Mountain shown in Figures 39 and 41. An excellent and readily accessible exposure of the basal zone occurs on the north side of a secondary road about 0.3 mile southeast of its junction with Pa. Route 507. The road junction is 1.9 miles north of Greentown, and the outcrop is at 41°20′50″N/75°18′15″W.

Good exposures of red siltstone and claystone occur in borrow pits near Lake Russell at 41°15′36″N/75°16′45″W and 41°15′05″N/75°18′10″W.

ORIGIN: The Poplar Gap and Packerton Members were deposited by shallow, braided streams flowing to the north and northwest from the Acadian mountains many miles to the southeast. This sediment-dispersal system passed approximately through Egypt Mills, and the system has been given that name. This sediment-dispersal system was distinctly different from the Port Jervis system, which deposited the Lackawaxen Member, and is distinct from those discussed previously by Sevon (1979). The Egypt Mills system presumably merged laterally with the Port Jervis system to the northeast and the Monroe system to the southwest. The areal persistence of the basal conglomeratic zone presumably corresponds to an extended period of coarse-grained sediment supply, during which time the braided streams wandered freely over the alluvial plain. The more sporadic distribution of the higher conglomeratic units suggests more variable sediment supplies.

Some flood-basin areas existed, particularly in the stratigraphically higher parts of the unit, and silts and clays were periodically deposited in these areas during flood events. Subsequent times of quiescence allowed growth of vegetation, a variety of small terrestrial invertebrates, oxidation of the fine-grained sediments to red, and at least incipient soil development.

MISCELLANEOUS: The red siltstone and claystone have been mined for use as random fill and are discussed in the section "Mineral Resources."

# SURFICIAL GEOLOGY

## INTRODUCTION

Much of the surface of Pike County has a thin cover of unconsolidated materials which are diverse in character and origin. All of these materials are related either to glaciation or to erosion and deposition subsequent to deglaciation.

All of extreme northeastern Pennsylvania has been glaciated at least three times (Sevon, 1974; Sevon and others, 1975), but evidence of only the last glaciation has been found in Pike County. This lack of evidence for older glaciations is not surprising because of (1) the lack of thick exposures of surficial material, and (2) the erosive capability of an ice sheet moving across the irregular topography of Pike County. The most recent glaciation in Pike County was late Wisconsinan (Woodfordian). The ice sheet probably entered the county at least 22,000 years ago, advanced to a maximum position more than 20 miles southwest of the county, and then melted, leaving the county free of ice probably as early as 18,500 years ago.

The amount of erosion by Woodfordian glacial ice is not known, but was certainly variable. Observations indicate that freshly exposed rock surfaces always are polished and have striations caused by the erosive base of the ice, but there is no indication as to how much rock was eroded by the ice. The thickness of unconsolidated fill in several of the valleys in the county attests to the fact that the valleys were eroded to considerable depth prior to being filled during and subsequent to glaciation.

A variety of unconsolidated surficial materials was deposited directly or indirectly by glacial ice. These deposits are discussed in the sections that follow, and their distribution is shown on Plate 2. Other surficial deposits do not owe their origin to glacial ice (e.g., shale-chip rubble and alluvium). These deposits are also described and shown on Plate 2.

## BEDROCK AND THIN OLEAN TILL, UNDIVIDED

DESCRIPTION: Much of the surface of Pike County is either exposed rock or rock covered by a layer of Olean Till less than 6 feet thick. The appearance of this surface varies from rock ledges (Figure 42) to relatively flat areas having a thin till cover (Figure 43).

Rock ledges of varying length and height are common throughout the county and are com-

Figure 42. Ledge of sandstone in the Delaware River Member of the Catskill Formation having a small accumulation of boulder debris at its base. The ledge occurs about 2.2 miles south-southeast of Twelvemile Pond at 41°08′37″N/75°04′49″W. The scale is divided in feet.

Figure 43. Olean Till overlying the striated surface of a sandstone in the Delaware River Member of the Catskill Formation. The till is less than 3 feet thick, and the area has been mapped as bedrock and thin Olean Till. The striae are oriented N23°E. The exposure is located 1.5 miles south-southeast of Gold Key Lake at 41°17′36″N/74°55′35″W. The scale is divided in inches and centimeters.

posed dominantly of sandstone, although some ledges are conglomerate, and some are capped with red siltstone or shale. Most ledges are 6 feet or less in height, but some exceed 20 feet. Almost all of the ledges have some boulder debris at their bases (Figure 42), and in a few places the areal extent of the debris is sufficient to warrant mapping as a boulder accumulation (see "Boulder Accumulations," p. 79–85). The boulder debris consists of angular fragments of rock derived mainly from adjacent outcrops and ranging in size from a few inches to a few feet. No size sorting is evident in these deposits. In some outcrops, the place of detachment for a large piece of rock is obvious (Figure 44). Rock ledges are most abundant on the steep slopes near the Delaware River northwest of Matamoras.

The uppermost surface of many rock ledges slopes at a low angle away from the ledge free face and is in many places relatively free of any cover, including vegetation (Figure 45). Some of these surfaces are relatively smooth and occasionally have glacial striations, but most have scattered broken rock and are sufficiently weathered so that striations have been obliterated. Subrounded to rounded glacial erratics locally occur on the bare surface (Figure 45). Sparse vegetation is established where roots can obtain hold, mainly along joints. The width of the exposed rock surface varies, but is generally 25 to 100 feet. The exposed rock eventually disappears beneath a cover of Olean Till, soil, and vegetation.

The till starts as a very thin coating and gradually becomes thicker with increasing distance

Figure 44. Ledge of sandstone in the Delaware River Member of the Catskill Formation and a large detached block. The width of the block is controlled by joints originally oriented N12°E. The ledge occurs along the south side of Snow Hill Road about 0.6 mile south of Twelvemile Pond at 41°10′05″N/75°05′38″W.

Figure 45. Glaciated bedding surface on sandstone of the Lackawaxen Member of the Catskill Formation. The surface is free of soil and till cover. The loose rocks on the surface were deposited by glacial ice. A rock ledge occurs to the left of the view in the photograph, and thin till occurs to the right of the view. The camera case rests on a glacial erratic. The exposure is located about 1.1 miles east of Cobey Pond at 41°34′04″N/75°02′52″W.

from the exposed rock. In some areas the till cover does not attain a thickness greater than 6 feet before another rock ledge is encountered. The general distribution of the Olean Till in relation to rock ledges and boulder accumulations is shown schematically in Figure 48.

Many areas in Pike County have few or no rock outcrops and are covered by less than 6 feet of till (Figure 43). These areas vary from gently rolling to moderately steep slopes. The character of the till is the same as that described in the following section "Olean Till," but the thickness is less than the arbitrary limit for mapping as a till deposit. Generally, the real thickness of the till is known only from well data or man-made exposures (Figure 46). The surfaces of some of these areas are relatively free of any rock debris, whereas some surfaces have moderate to abundant amounts of rock debris (Figure 47).

THICKNESS: By definition the thickness of the Olean Till in this unit is less than 6 feet.

CONTACTS: The contact of the Olean Till with the underlying rock in this unit is sharp. The contacts of this unit with adjacent mapped units are generally arbitrary and based on field and/or aerial-photograph interpretations. Contacts with adjacent areas of Olean Till are based on estimated changes in thickness of till from less than 6 feet to greater than 6 feet and probably vary in accuracy from place to place.

EXPOSURES: Natural exposures of rock ledges abound in this county, and many are referred to in the figures in this section and those in the section "Surface Bedrock Stratigraphy." Exposures of the Olean Till less than 6 feet thick overlying rock are mainly temporary, man-made exposures, although those shown in Figures 43 and 46 may still be in existence.

ORIGIN: This mapped unit resulted from glacial erosion of the area by Woodfordian ice and deposition of till of varying thickness by that ice during advancing and wasting phases. Subsequent erosion has little modified the surface.

Figure 46. Olean Till overlying sandstone of the Towamensing Member of the Catskill Formation. The till is less than 6 feet thick, and the area is mapped as bedrock and thin Olean Till. The exposure is located 1.1 miles north-northwest of Winona Falls at 41°07′26″N/75°03′27″W.

## OLEAN TILL

INTRODUCTION: Modern mapping of unconsolidated surficial deposits of glacial origin began in northeastern Pennsylvania with the work of Epstein (1969). Neither Epstein, who used genetic mapping units and terminology, nor subsequent workers (e.g., Berg, 1975; Sevon, 1975a, b, c; Berg and others, 1977; and Sevon and Berg, 1978), who used time-stratigraphic-genetic mapping units and terminology, have applied formal rock-stratigraphic terminology to glacial till units in northeastern Pennsylvania. Crowl and Sevon (1980) proposed formal acceptance of the established name Olean for drift of young age in eastern Pennsylvania. This includes all drift having similar age-related characteristics (degree of soil development, extent of erosion, and position north of the Woodfordian border [Crowl and

Figure 47. Slope having some undisturbed outcrops of sandstone of the Delaware River Member of the Catskill Formation (in the background), some glacially transported sandstone blocks from local outcrops, and a thin covering of Olean Till. The thickness of the till increases toward the viewer. The slope is located on the east side of Pa. Route 402, immediately north of Pickerel Lake at 41°15′35″N/75°04′44″W. The scale is divided in feet.

Sevon, 1980, Plate 1; the "terminal moraine" of Lewis, 1884]), and thus includes all of the till in Pike County.

Sevon and Crowl studied till outcrops throughout the county and analyzed 81 till samples in order to ascertain whether or not more than one distinct till exists in Pike County and whether or not local names should be applied. They concluded that, although outcrop appearance and pebble-lithology analysis support the existence of two distinct tills, this distinctiveness is not supported by textural data. Therefore, the authors use the name Olean for all of the till in Pike County, but discuss the areal variations.

The Olean Till shown on Plate 2 was mapped where it was considered to be 6 feet or more thick. Areas having less than 6 feet of till are discussed in the preceding section "Bedrock and Thin Olean Till, Undivided."

FIELD OCCURRENCE: Deposits of the Olean Till greater than 6 feet in thickness occur mainly on lower slopes and in valley bottoms throughout the county. This distribution is most striking in the southeastern half of the county, where a pronounced linear pattern occurs (Plate 2).

Uplands and hilltops throughout the county are either free of till cover or have only a thin (less than 6 feet) till cover (see "Bedrock and Thin Olean Till, Undivided" on Plate 2). Locally, the Olean Till is distributed in the manner shown in Figure 48. The upper surface of rock near a rock ledge is free of till. Downslope from the ledge, till occurs, and it increases in thickness

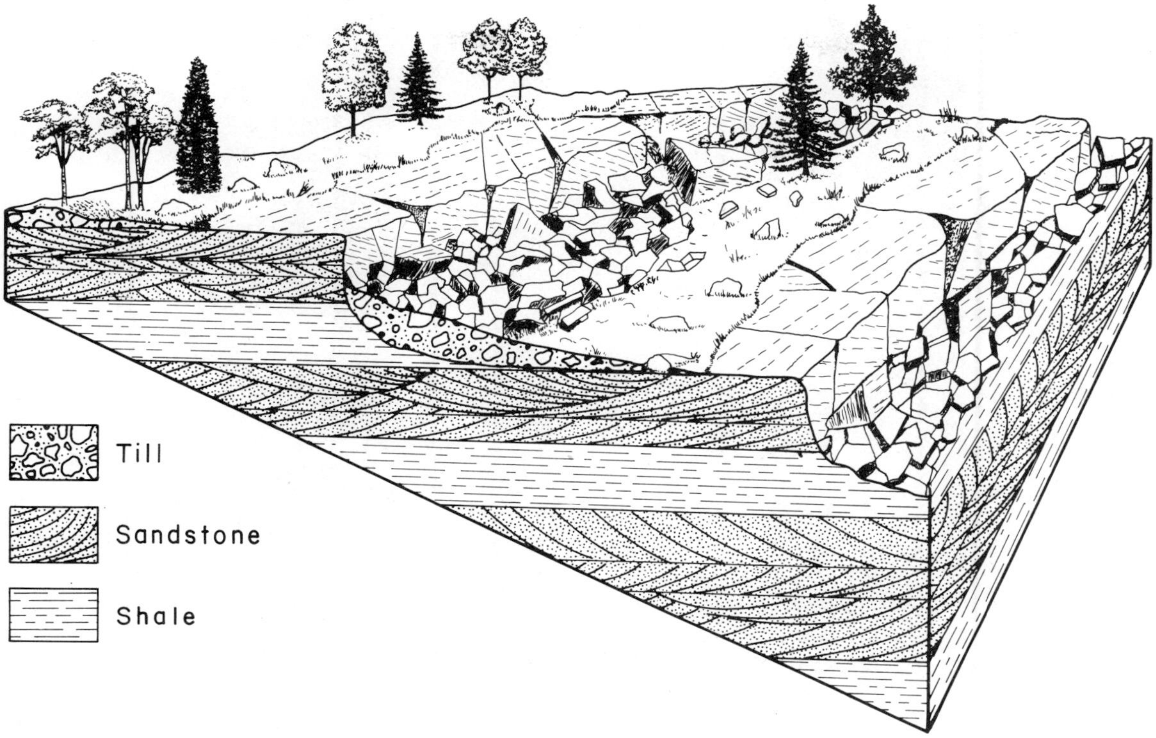


Figure 48. Typical distribution of Olean Till in relation to rock ledges and boulder accumulations.

until the base of another ledge is encountered. Boulder accumulations occur at the bases of ledges.

DESCRIPTION: The Olean Till is an unsorted mixture of boulders, cobbles, pebbles, sand, silt, and clay (Figure 49). Data from 78 samples collected throughout the county and three samples from just outside the county (Figure 50) indicate that the average till is 32 percent pebbles and 68 percent sand-silt-clay. The relative concentrations of gravel, sand, and silt and clay are shown in Figure 51, and the data are presented in Appendix 1. Size distribution values determined in the field or from larger samples would probably indicate less sand-silt-clay because of the presence of cobbles and boulders not collected for laboratory analysis. Proportions of the finer grained part of the till are shown in Figure 52, and the data are presented in Appendix 2. According to these data, the average till has 51 percent sand, 29 percent silt, and 20 percent clay. On the basis of the above textural data, the Olean Till in Pike County could be called a muddy, gravelly sand. Such terminology does not adhere totally to any specific textural classification, but approximates that of Folk (1954, p. 346–351). The till is generally not calcareous. The depth of oxidation is not definitely known, but is probably less than 12 feet. Color is variable and is discussed below. Cobbles and boulders composed of fine-grained rocks such as very fine grained sandstone and siltstone are commonly striated (Figure 53), but cobbles and boulders composed of coarser grained rocks are generally not striated.

Igneous and metamorphic rock clasts derived from sources in Canada or the Adirondack Mountains of New York State are extremely rare in Pike County. One boulder of granitic gneiss was found in till along Masthope Creek, and two

Figure 49. Exposure of Olean Till located in a roadcut 1.7 miles northeast of Hawley at 41°29′20″N/ 75°09′02″W. The scale is divided in feet.

cobbles of gneiss were found in the Delaware River valley (Crowl, 1971, p. 8).

Rounded boulders and cobbles of white quartzitic sandstone and conglomerate, presumably derived from the Shawangunk Formation in New York or New Jersey, are common in the eastern half of the county east of a line extending approximately southwest from Shohola past Porters Lake into Monroe County. These erratics are easily recognized because of their white color and commonly pitted surfaces. Their occurrence and distribution was first noted by White (1882, p. 42–43). The distribution of Shawangunk rocks is problematic because their source is to the northeast and east of Pike County, but striae indicate a consistent south-southwest flow of ice across the county except in the southeastern part of the county where striae immediately adjacent to the Delaware River valley indicate a southwest flow of ice. The only explanation presented to date is that of distinct but overlapping ice tongues (Sevon and others, 1975).

Cobbles and boulders of gray sandstone dominate the larger clasts occurring in the till throughout the county. Large clasts of red sandstone or siltstone occur but are not abundant. There is a definite relationship of till-clast lithology to underlying bedrock. Because the character of the Olean Till is strongly influenced by underlying bedrock and some variations in the till do occur, the remainder of the discussion on the physical character of the till will be by the areas designated in Figure 50.

Area A (Figure 50) is underlain by rocks of the Catskill Formation, and the character of the Olean Till in this area has the greatest variation. The average till is composed of 69 percent sand-silt-clay and 31 percent gravel, but the range of component values, 93 to 40 percent for sand-silt-clay and 60 to 7 percent for gravel, is large (Appendix 1). The fine-grained part of the average till comprises 52 percent sand, 28 percent silt, and 20 percent clay, but the range of values, 63 to 38 percent for sand, 40 to 18 percent for silt, and

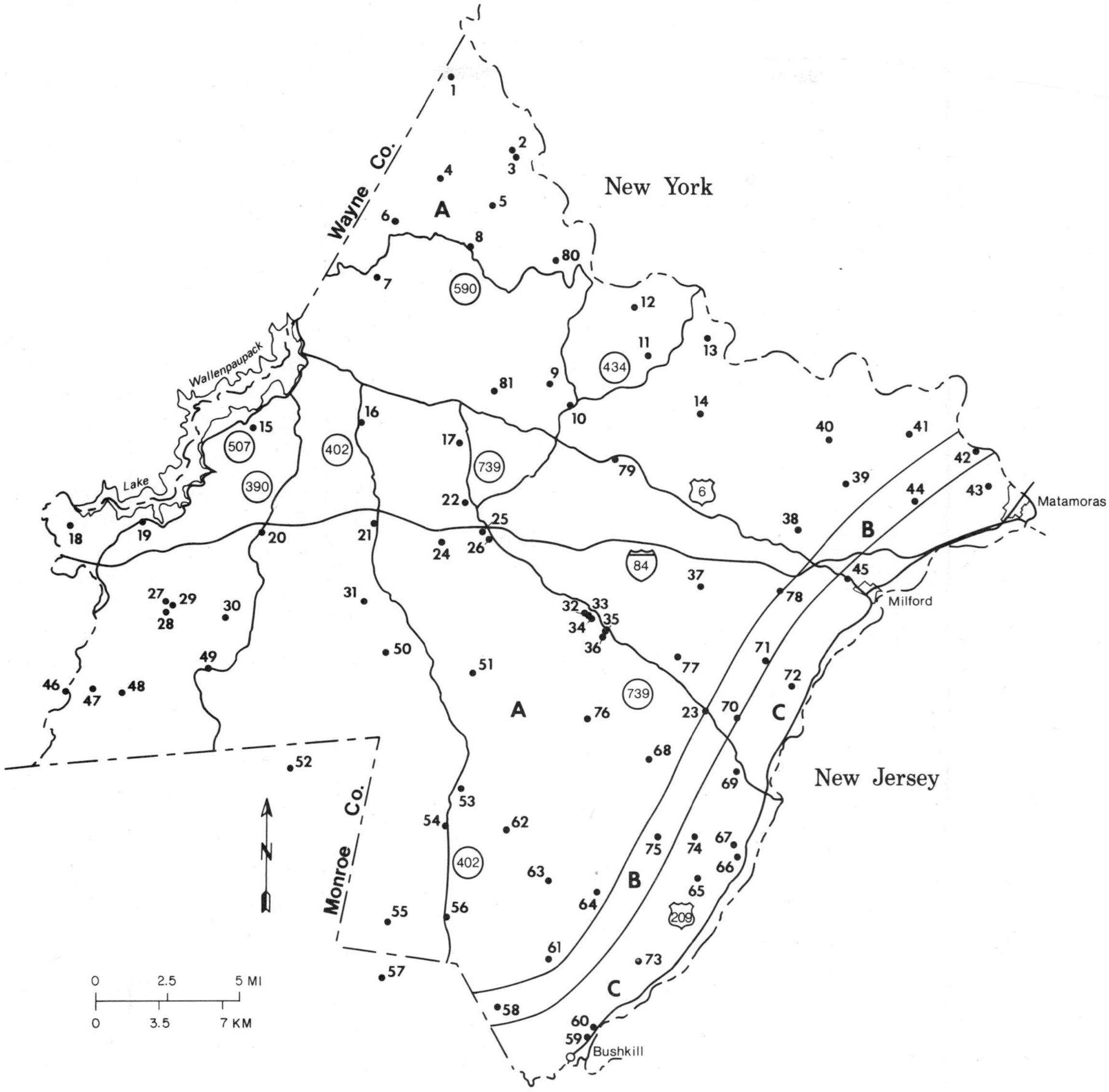


Figure 50. Location of 81 samples of Olean Till collected for analysis and three areal subdivisions (A, B, and C) discussed in the text.

36 to 9 percent for clay, is large (Appendix 2). This large range of grain-size values reflects the diversity of lithologies present in the underlying Catskill Formation and the close relationship between bedrock and till.

Pebble-lithology identifications and counts were done for all till samples, and the results are presented graphically in Figure 54. There is considerable variation in till-pebble composition within Area A, as can be seen by comparing the histograms of samples 1, 36, and 41 (Figure 54). Therefore, although the summary histograms in Figure 54 indicate significant differences in till-pebble composition between areas, a specific sample may not be diagnostic of a particular area. In general, however, tills in Area A contain more

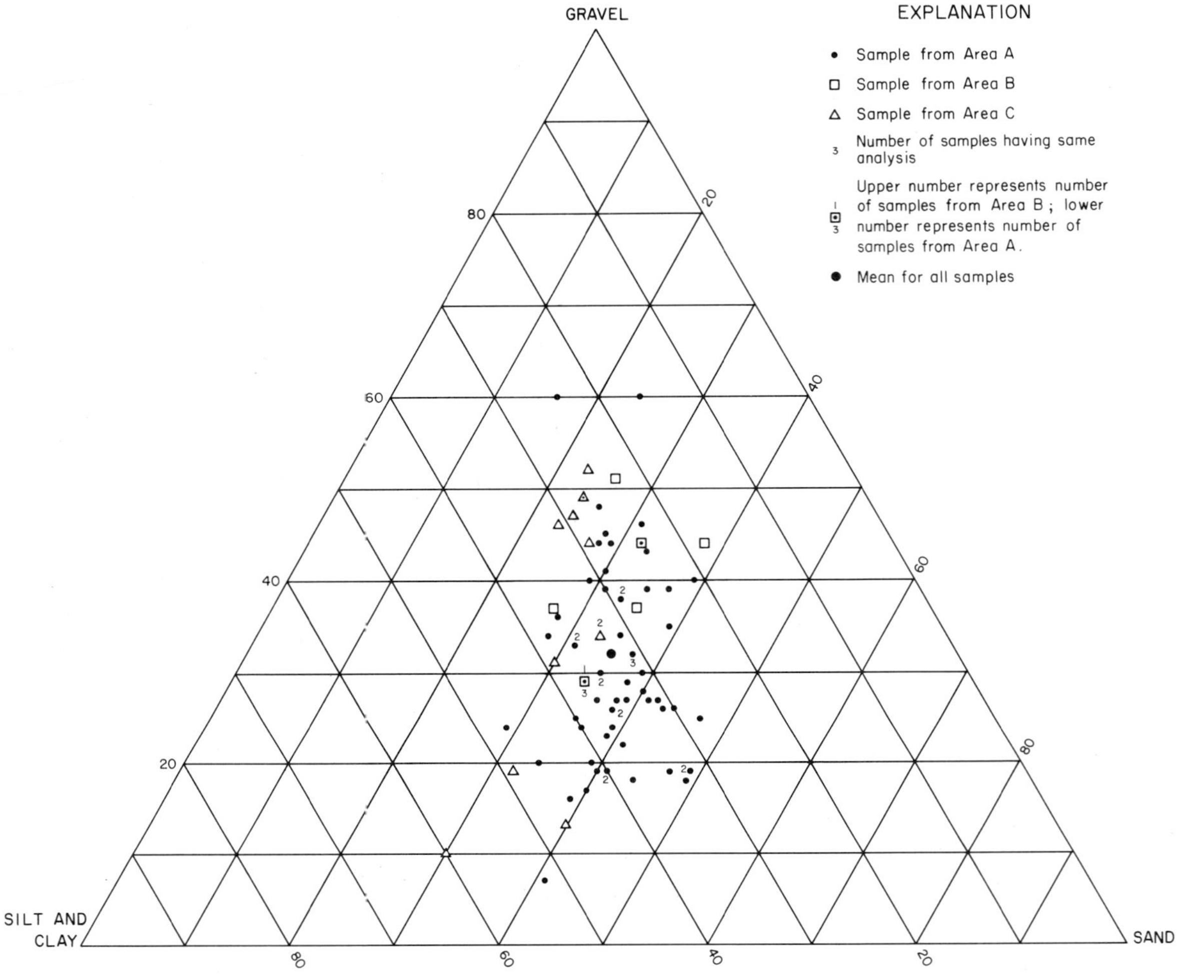


Figure 51. Triangular diagram of the gravel, sand, and silt-clay distribution of 81 samples of Olean Till. The sample localities are shown in Figure 50.

sandstone pebbles, a wide range of sandstone grain sizes, and some red pebbles, commonly claystone or siltstone. The dominance of sandstone as a component is also reflected by the large number of sandstone cobbles and boulders present within and on the surface of the Olean Till. Grain components in the 2- to 4-mm fraction usually have the same compositions as those in the coarser fraction, although there is a tendency for the quantity of red claystone and siltstone in the 2- to 4-mm fraction to be disproportionately larger than the quantity in the coarser fraction. In sizes smaller than 2 mm, the quantity of rock fragments decreases and the quantity of quartz grains increases progressively, and quartz is dominant in all grain sizes smaller than 0.5 mm.

The color of the Olean Till in Area A is variable and is influenced by the red beds of the Catskill Formation. Over half of the tills are reddish brown (5YR5/3–4/4–4/3),[1] about one quarter are dark brown (7.5YR4/4) or variations of brown (10YR5/3–4/3 and 7.5YR5/4–5/6), and

[1]Color designations for surficial units are from *Munsell Soil Color Charts*, Munsell Color (1975).

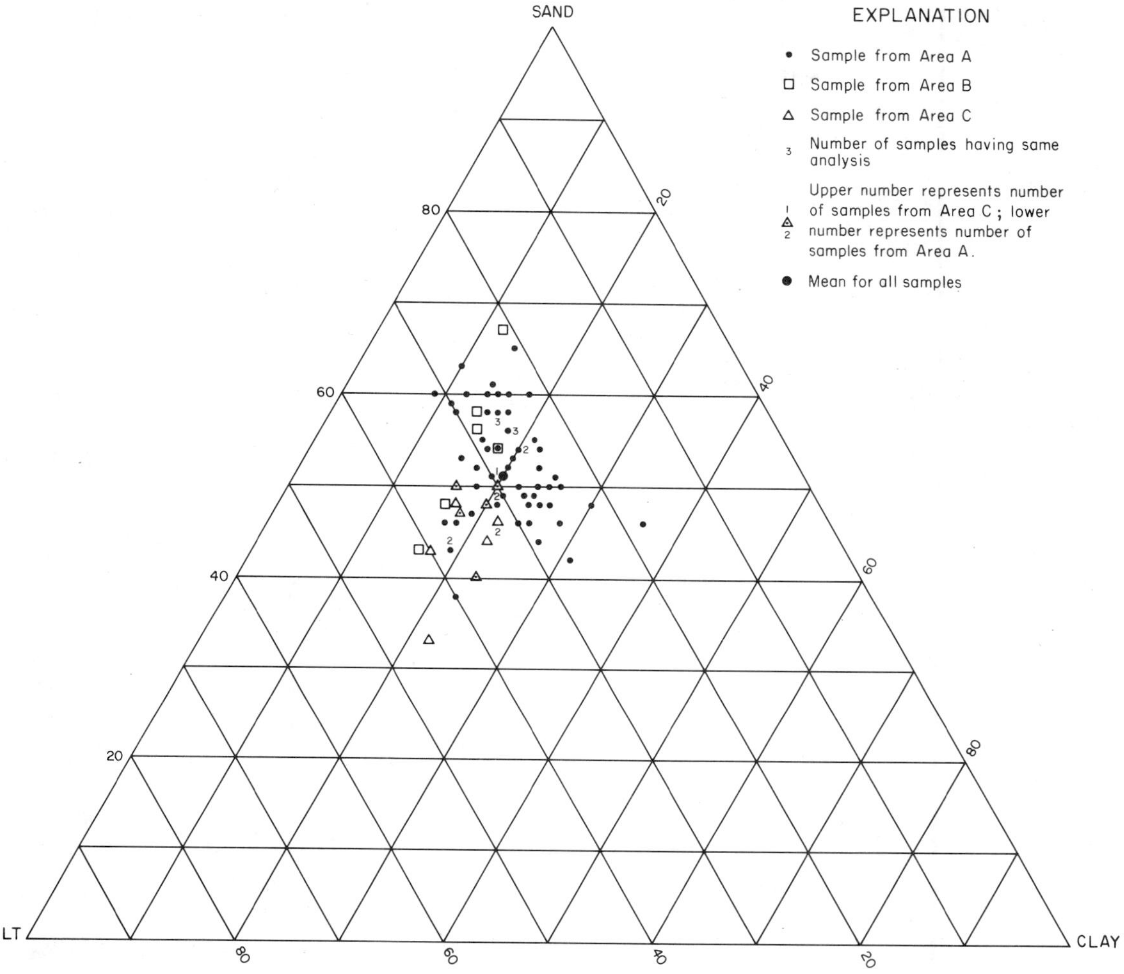


Figure 52. Triangular diagram of the sand, silt, and clay distribution of 81 samples of Olean Till. The sample localities are shown in Figure 50.

the remainder are yellowish brown (10YR5/4–4/4–4/3).

The Olean Till in Area A is generally compact in fresh exposures but has moderate to poor cohesion and breaks up easily when handled. Boulders and cobbles of gray sandstone are common within and on the surface of the till. In areas where large numbers of boulders occur, they have been mapped as "Boulder Accumulations" (Plate 2; "Boulder Accumulations," p. 79–85). An exposure of the Olean Till presumed typical of Area A is described in Section 1 of Appendix 3. The soil profile present in this exposure is readily recognizable but not well developed. Levine (1981) and Levine and Ciolkosz (1983) indicated that relatively few physical changes occurred during the soil-profile development in till of Woodfordian age. Descriptions and laboratory data for several soil profiles developed in the Olean Till are presented by Cunningham and others (1972, p. 596–599, 602–603, 612–613, 616–619). Additional soil information is given in the Pike County soil survey (Taylor, 1969).

Area B (Figure 50) is underlain by rocks of the Trimmers Rock Formation. The average till is composed of 60 percent sand-silt-clay and 40 per-

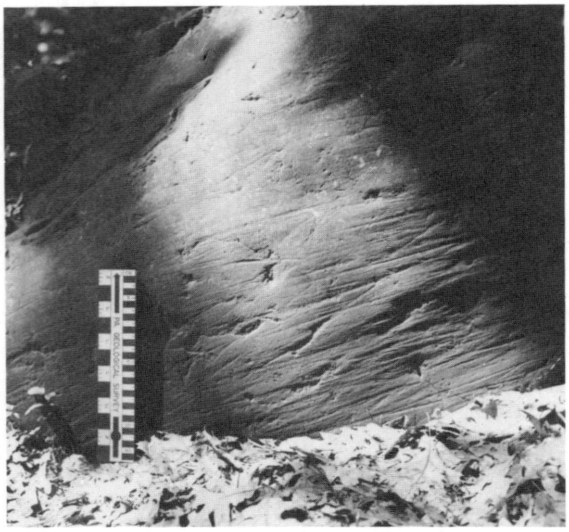

Figure 53. Boulder of dark-gray siltstone having well-developed surface striations. The boulder is located on the south side of a road 0.3 mile northwest of Millrift at 41°24′48″N/74°44′47″W. The scale is divided in inches and centimeters.

cent pebbles, and the range of values is relatively small (Appendix 1). The fine-grained part of the average till comprises 54 percent sand, 31 percent silt, and 15 percent clay, and has a relatively small range of values (Appendix 2). Although the relatively small range of values may be partly due to the small number of samples, the range presumably also reflects the relatively uniform character of the underlying bedrock.

Pebble-lithology identifications for the six samples collected in Area B are presented in Figure 54. Sample 58 is anomalous, being more typical of till that has a strong Catskill bedrock influence. The remainder of the samples are made up of dominantly gray sandstone and siltstone. Red lithologies are very minor to absent, and gray sandstone dominates surface boulders and cobbles.

The color of the Olean Till in Area B is relatively uniform and is brown (10YR5/3–4/3). Color identifications for Area B were all made on oxidized till; unoxidized till may have a different chroma.

The Olean Till in Area B is generally compact in fresh exposure, but has moderate to poor cohesion and breaks up readily when handled. Boulders and cobbles of gray sandstone are common on the surface of the till but are less abundant than in Area A.

Area C (Figure 50) is underlain by rocks of the Mahantango Formation and has the most uniform till. The Olean Till occurring in this area contrasts most to that occurring in Area A. The average till is composed of 66 percent sand-silt-clay and 34 percent gravel, but the range of component values, 90 to 48 percent for sand-silt-clay and 52 to 10 percent for gravel, is almost as large as that in Area A (Appendix 1). The fine-grained part of the average till comprises 45 percent sand, 35 percent silt, and 20 percent clay, and has the smallest range of values for any of the three areas (Appendix 2). These latter values appear to reflect the uniform siltstone lithology of the Mahantango Formation.

Pebble-lithology identifications for the samples collected in Area C (Figure 54) indicate a very uniform clast composition comprising mainly gray siltstone and some gray sandstone. This uniform clast composition is an excellent indicator of the strong influence that the underlying bedrock has on the character of the overlying till. This influence was documented elsewhere in northeastern Pennsylvania by Epstein (1969, p. 55–59).

The Olean Till in Area C in the oxidized zone is yellowish brown (10YR5/4) and slight variations of that color. The Olean Till in the only unoxidized exposure in Area C (Horizon D, Section 2, Appendix 3) is dark gray (7.5YR4/0).

The till in Area C is moderately compact and cohesive where oxidized and very compact and cohesive where unoxidized. Till in the single unoxidized exposure is slightly calcareous, and rare limestone pebbles occur in surface float, although none were present in the analyzed sample. Section 2 in Appendix 3 is a description of the best exposure of the Olean Till in Area C. Descrip-

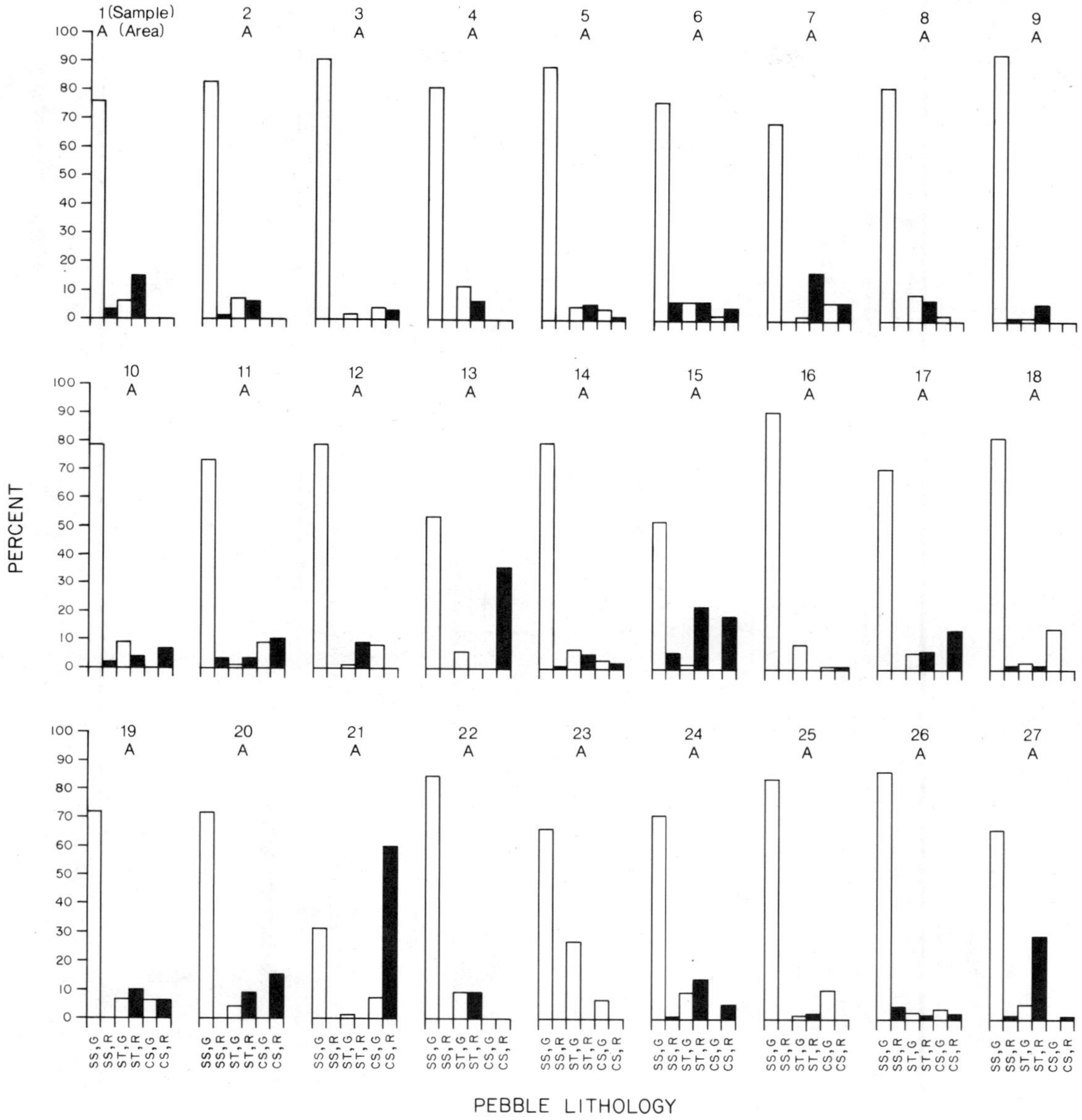


Figure 54. Histograms showing the lithology of pebbles larger than 4 mm in diameter found in 81 samples of Olean Till whose locations are shown in Figure 50. Each histogram is identified by sample number and area. The major lithologic components shown are as follows: SS, sandstone; ST, siltstone; and CS, claystone. Minor lithologic components include the following: CGL, conglomerate; QZ, quartz; and QZT, quartzite. Modifiers listed after some of the lithologic components are as follows: G, gray; R, red; CG, coarse grained; MG, medium grained; and FG, fine grained. Red lithologies are shown as solid bars on the histograms, and trace amounts are represented by the letter T. The histograms for individual samples show only the major lithologic components, and the summary histograms show all identified lithologies.

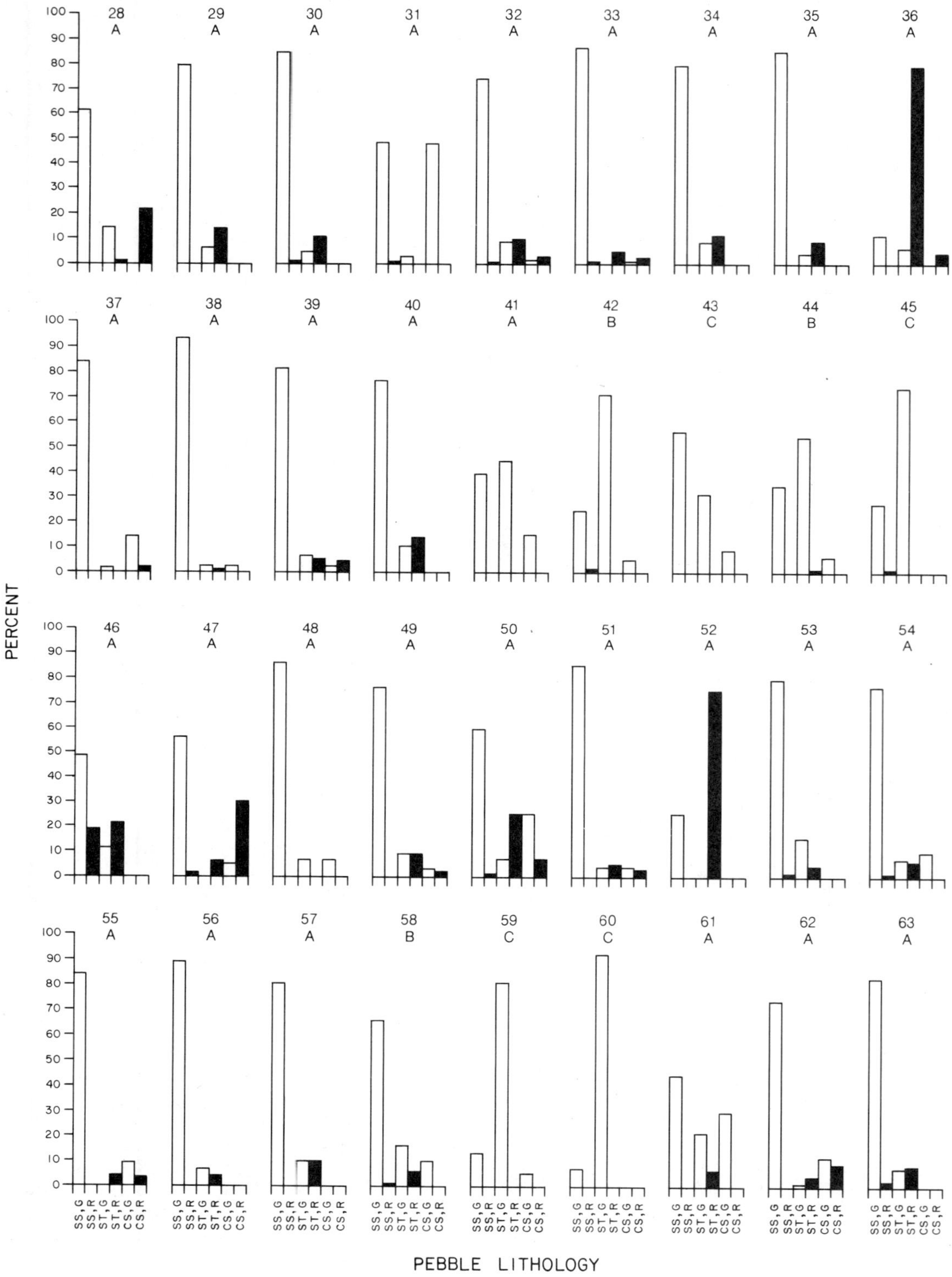


Figure 54. *(Continued).*

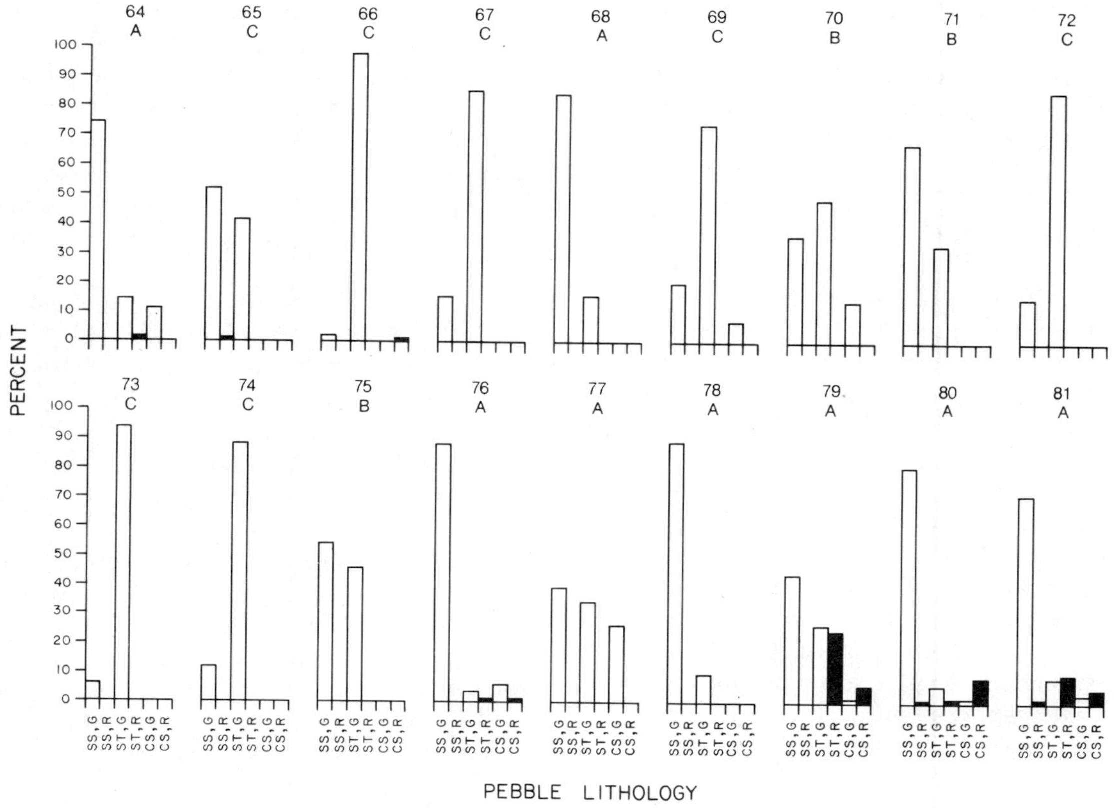


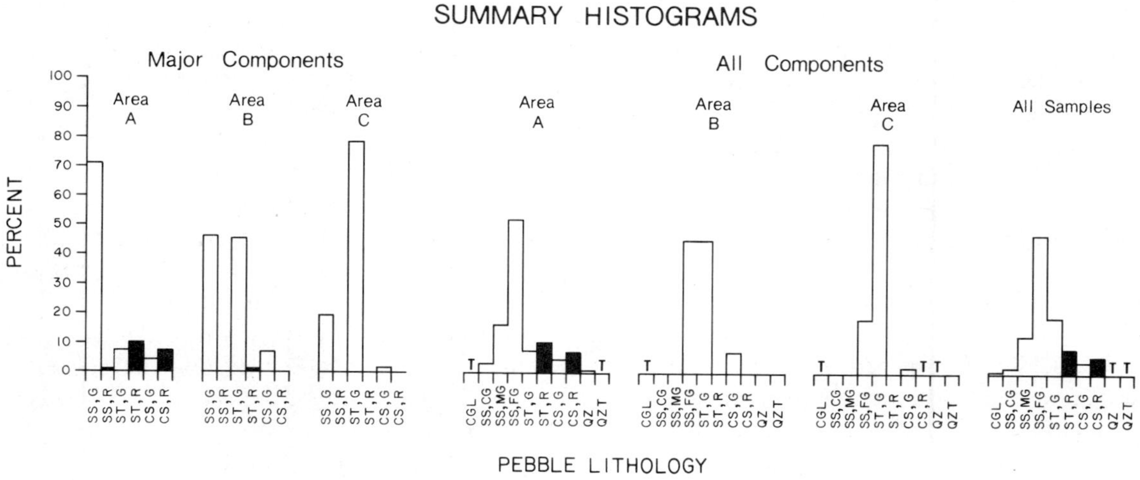


Figure 54. *(Continued).*

tions of soil profiles developed on till characteristic of Area C are presented in Cunningham and others (1972, p. 604–607), and other soil information is presented in the Pike County soil survey (Taylor, 1969).

THICKNESS: The thickness of the Olean Till mapped in Pike County ranges from the lower defined limit of 6 feet to a presumed maximum of over 200 feet in a water well near Greeley. Exposures seldom show more than a few feet of till, but water well records indicate moderately thick till in the county. A mean till thickness of 88 feet (standard deviation, 73 feet) was obtained from thickness data in records for 250 water wells, in which unconsolidated material was not specified as sand and gravel. A mean till thickness of 94 feet (standard deviation, 42 feet) was obtained from thickness data in records for 55 water wells that were definitely drilled into till. In general, thin till or no till occurs on the uplands, till thickness increases progressively downslope, and till thickness is greatest in valley bottoms. There are numerous exceptions to this generality.

CONTACTS: The Olean Till probably directly overlies bedrock everywhere. The rock surface beneath the till is irregular, generally very smooth to polished, and marked by numerous scratches and small grooves (striations) (Figure 55). The smooth rock surface and the striations are commonly destroyed by prolonged exposure to normal weathering conditions.

The upper surface of the Olean Till is either the land surface itself or a contact with other surficial deposits, which in places overlie the till. The contact with any of the overlying deposits ranges from sharp to transitional, depending on how much mixing occurred during the deposition of the overlying material.

EXPOSURES: Temporary exposures occur throughout Pike County in excavations, roadcuts, and ditches. Self-perpetuating exposures occur in the faces of slope failure along Masthope Creek (Figure 56) and near Bushkill (Section 2, Appendix 3).

ORIGIN: The Olean Till resulted from continental glaciation during the Woodfordian Substage of the Wisconsinan Stage, between about 22,000 and 19,000 years ago. During that time, an ice sheet at least 1,000 feet thick near its margin advanced slowly from the north across parts of eastern Pennsylvania to a maximum position southwest of Pike County (Crowl and Sevon, 1980).

Ice flow was slightly west of south in the northern and western parts of the county and almost due southwest in the southeastern part of the county, where the direction of ice flow was controlled by the Delaware River valley and the Kittatinny Mountain on the east side of the river. Directions of ice flow are indicated by orientations of striae shown on Plate 2.

As the ice advanced, erosion by abrasion and plucking occurred. Fine-grained debris entrained in basal ice abraded and polished the rock surfaces, and larger rock debris scratched the rock surfaces to produce striae. Much of the material eroded by the ice was deposited as till. Deposition occurred in two ways. As the ice advanced, some debris was plastered onto the underlying surface as an accretionary deposit called lodgement till. After ice motion stopped and melting commenced, some debris accumulated in situ, without transport by running water, as melt-out till (ablation till). Some of the melt-out debris was resedimented locally.

## OLEAN ICE-CONTACT STRATIFIED SAND AND GRAVEL

INTRODUCTION: Deposits of sand and gravel vary in size, quality, and origin throughout Pike County. Discussed in this section are the deposits that were deposited by water in close proximity to glacial ice. These deposits are characteristically quite variable in size of constituent materials, lateral and vertical continuity of individual beds, thickness, and areal dimension. The ice-contact deposits are an important economic resource in Pike County and are discussed in detail in the section "Mineral Resources."

Figure 55. Striations scratched into gray sandstone of the Lackawaxen Member of the Catskill Formation. The striations are oriented N50°E. The exposure is of the upper surface of a sandstone exposed in a roadcut on the south side of U.S. Route 6 just east of its junction with Pa. Route 507 south of Wilsonville at 41°26′36″N/75°10′38″W. The scale is divided in inches and centimeters.

FIELD OCCURRENCE: Most of the deposits of Olean ice-contact sand and gravel occur either in valley bottoms or on valley sides. A few deposits occur on relatively flat upland sites. Most of the valley-bottom deposits have an irregular surface topography of rounded hills and undrained depressions, and a relief of a few feet to a few tens of feet. Some of these deposits cover most of the valley floor. Other deposits occur as isolated hills within otherwise flat-floored valleys. These deposits are generally designated as kames.

Numerous deposits occur along valley walls. These deposits may be topographically distinct from the bedrock valley wall, or they may form relatively smooth slopes that blend into the bedrock slopes. These deposits are designated as valley-side kames. Some deposits along the sides of the Delaware River valley have a downstream-sloping, relatively flat to undulating upper surface, a steep upstream terminus, a gentle to moderately steep river-facing slope, and a downstream end that either blends into the floodplain surface or terminates at the upstream end of a similar

A

B

Figure 56. Olean Till exposed in a slump scar on the north side of Masthope Creek about 0.9 mile northwest of Masthope at 41°32′46″N/75°02′24″W.
A. Closeup of the west end of the scar.
B. Distant view of the east end of the scar.

deposit. These deposits, called kame terraces, are discussed in more detail by Crowl (1971).

DESCRIPTION: Deposits of Olean ice-contact stratified sand and gravel comprise a heterogeneous assemblage of boulders, cobbles, gravel, sand, silt, and some clay sorted into beds that have sharp boundaries and marked contrasts in sediment size (Figure 57). Most deposits are mixed sand and gravel, but a few deposits are almost entirely sand.

The deposits of ice-contact stratified mixed sand and gravel have extremes in grain size ranging from interstitial clay to boulders up to 6 feet in diameter (Figure 58). Boulders up to 6 feet in diameter are uncommon, boulders up to 3 feet are more common, and those up to 1.5 feet are most common. The boulders, composed of partly rounded, gray sandstone, occur randomly within a deposit and constitute only a small percentage of the deposit. Well-rounded clasts less than 25 cm in diameter are the most abundant coarse clasts in all of the deposits of mixed sand and gravel, and, in many of the deposits, the largest part of this size range is between 7 and 15 cm. Individual layers within a specific deposit may have clasts of approximately the same size, but grain size and degree of sorting normally vary greatly from one layer to an adjacent layer (Figure 59). A few deposits are composed of relatively uniform accumulations of similar-sized materials (Figure 60).

The orientation and thickness of beds in the deposits of sand and gravel are quite variable but generally well defined. Individual beds range in thickness from less than an inch to more than a few feet, but commonly are less than 1.5 feet thick. Adjacent beds may have markedly different thicknesses. The dip of the beds in any deposit may range from horizontal to vertical, and the direction of dip may be variable. In many places, a sequence of several beds which dip in

Figure 57. Deposit of Olean ice-contact stratified sand and gravel in one of the Pike County Sand and Gravel Company quarries on the west side of Blooming Grove Creek, about 1 mile northwest of Lake Giles at 41°23′13″N/75°06′40″W. Note the sharp boundaries between beds containing sediment of contrasting size.

one direction is truncated by an erosional contact, and the overlying beds dip in the opposite direction.

The composition of the gravel-sized material in the ice-contact deposits is variable and related to the underlying bedrock. A summary of pebble lithologies for gravel samples collected throughout Pike County is presented in Figure 61. Figure 62 shows the locations of the sampling sites. The dominant pebble lithology in all of the samples is very fine to fine-grained, gray sandstone. Figure 61A shows a higher percentage of gray siltstone and claystone clasts than any of the other sample groups; this reflects the lithologies of the underlying Trimmers Rock and Mahantango Formations. In similar fashion, quantities of red lithologies are shown in Figure 61C and D and reflect the presence of red beds in the underlying rocks. More coarse-grained material is shown in Figure 61D than in any of the other suites, which again reflects the nature of the underlying bedrock. The analyses do not include the composition of material coarser than about 8 cm, but data from field observations indicate that the larger clasts are mainly sandstone and some other compositions, such as red siltstone.

A few deposits in Pike County are composed entirely of sand (Figure 63) or mainly of sand covered by a thin gravel cap. These deposits are characterized by well-defined beds and in many places have extensive development of ripples within individual layers (Figure 64). The sand is generally moderately well sorted and uniform in size distribution within individual beds, and commonly very little variation in overall size occurs between different beds. Some deposits contain

Figure 58. Large boulders in a deposit of Olean ice-contact stratified sand and gravel in an old borrow area 1 mile southeast of Germantown at 41°24′28″N/75°03′46″W.

thin layers of silt or clay, which are cohesive and usually slightly red in color (Figure 65). These finer grained, cohesive beds make the deposit more resistant to erosion than deposits composed mainly of unconsolidated sand. Bedding in these deposits is generally horizontal to subhorizontal. Steeply dipping beds and variable orientations of bed dips are uncommon.

In Table 5, data are presented based on visual estimates of the composition of sand in samples from deposits of ice-contact sand and gravel in Pike County. These data indicate a pronounced relationship between sand composition and underlying bedrock. For example, note the contrast in the quantity of red rock fragments between samples overlying nonred-bed rock units (Table 5, columns A and B) and those overlying red-bed rock units (Table 5, columns C and D). In all samples the quantity of quartz increases and the quantity of rock fragments decreases as grain size decreases.

THICKNESS: The thickness of deposits of Olean ice-contact stratified sand and gravel is variable and in general is not well known. Several deposits, which have been quarried or exposed by erosion, indicate a thickness in excess of 100 feet. The thickness of most deposits is difficult to estimate because of the lack of information about the extent of the deposit below the surface of the ground, but an average thickness of 30 to 60 feet seems reasonable for deposits of sand and gravel in Pike County.

CONTACTS: The deposits of Olean ice-contact stratified sand and gravel presumably overlie bedrock or till, but the base of these deposits is seldom exposed. The deposits are generally not overlain by any other deposits.

EXPOSURES: The best exposures of Olean ice-contact sand and gravel in Pike County occur in numerous quarry operations (particularly in the quarries operated by Pike County Sand and

Figure 59. Variations in grain size between adjacent layers in a deposit of Olean ice-contact stratified sand and gravel in one of the Pike County Sand and Gravel Company quarries mentioned in Figure 57. The scale is divided in inches and centimeters.

Figure 60. Deposit of Olean ice-contact stratified gravel exposed in a small quarry on the east side of Blooming Grove Creek about 0.2 mile downstream from Fourmile Pond at 41°26′14″N/ 75°06′14″W. The shovel handle is 27.5 inches long.

Gravel Company along Blooming Grove Creek), several small deposits along Blooming Grove Creek between U.S. Route 6 and the Lackawaxen River, and some deposits near Wallenpaupack Creek near Greentown. Locations of several exposures are given in Figures 57 to 60 and 63 to 65.

ORIGIN: Olean ice-contact stratified sand and gravel was deposited in close proximity to Woodfordian (late Wisconsinan) glacial ice during the melting phase of that glaciation. The materials were deposited by swift water that flowed off the ice into an opening within or adjacent to the ice. The deposits that formed in pockets within the ice are called kames and may have fine-grained sediments indicative of standing water (Figure 65). After the ice melted completely, these deposits became isolated small hills in valley bottoms. The water that flowed off the glacial ice into open areas between the ice and valley walls deposited valley-side kames or kame terraces.

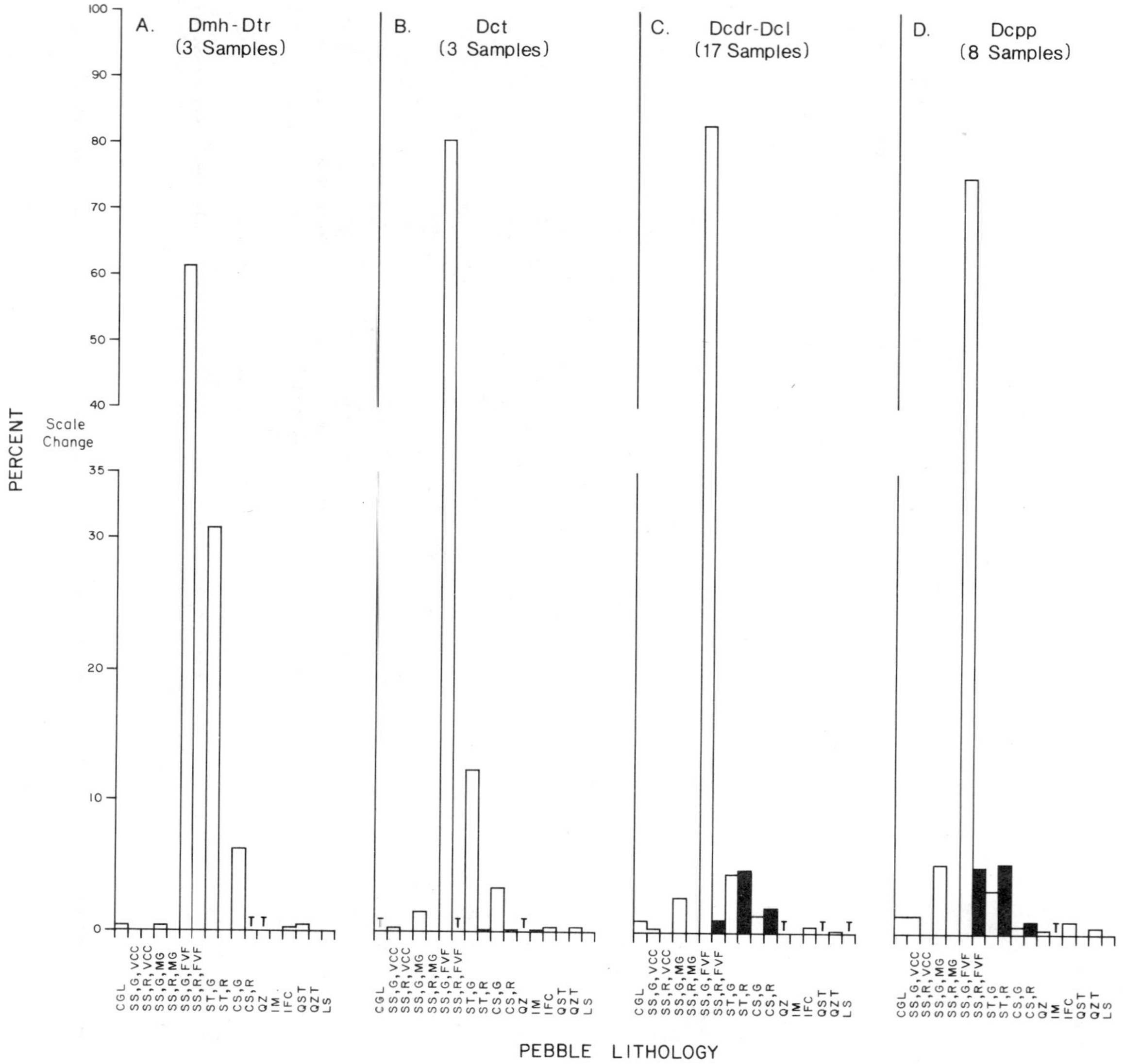


Figure 61. Histograms showing the lithology of gravel-sized clasts greater than 0.5 cm in analyzed sand and gravel deposits. Note the change in the percentage scale. The histograms are separated by underlying bedrock: Dmh-Dtr, Mahantango and Trimmers Rock Formations; Dct, Towamensing Member of the Catskill Formation; Dcpp, Poplar Gap and Packerton Members of the Catskill Formation. Lithology is as follows: CGL, conglomerate; SS, sandstone; ST, siltstone; CS, claystone; QZ, quartz; IM, igneous and metamorphic; IFC, intraformational conglomerate; QST, quartzitic siltstone; QZT, quartzite; LS, limestone. Modifiers include the following: G, gray; R, red; VCC, very coarse to coarse grained; MG, medium grained; and FVF, fine to very fine grained. Red lithologies are shown as solid bars, and trace amounts are represented by the letter T.

Valley-side kames generally have an irregular surface topography and variable bed orientations and sizes of material (Figures 57 and 59). Kame terraces have a steep upstream margin and a surface that slopes downstream. These terraces indicate deposition in an area between the ice and valley wall that had an outlet at the downstream end.

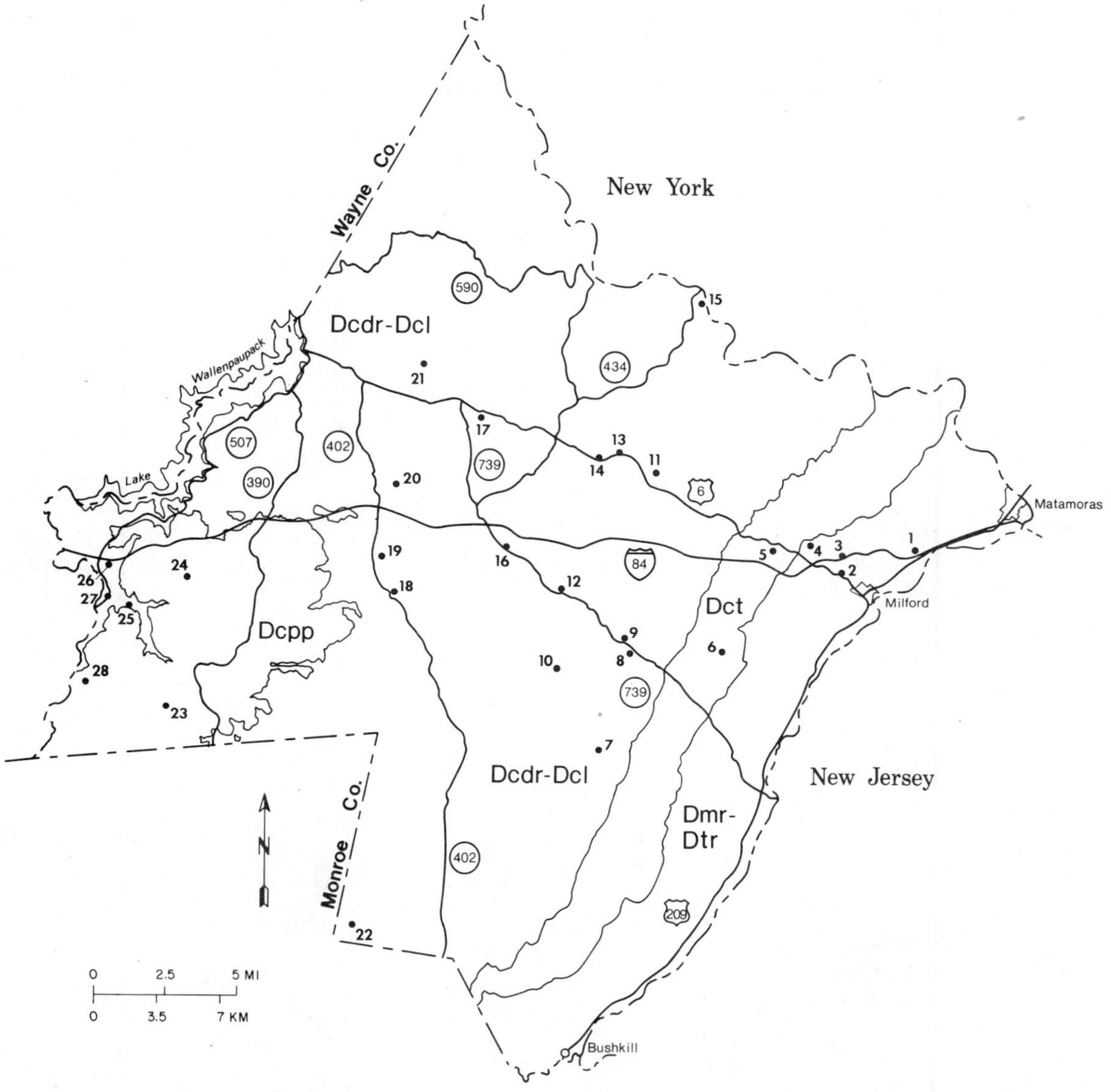


Figure 62. Location of sand and gravel sampling sites in Pike County. Dmr-Dtr, Marcellus through Trimmers Rock Formations; Dct, Towamensing Member of the Catskill Formation; Dcdr-Dcl, Delaware River and Lackawaxen Members of the Catskill Formation; Dcpp, Poplar Gap and Packerton Members of the Catskill Formation.

Evidence of proximity to glacial ice is commonly present in these deposits. Figure 66 shows two small normal faults that probably developed when ice surrounding a kame deposit melted and the materials in the deposit collapsed. Folding of stratified layers is seen in some exposures and indicates overriding of a deposit by glacial ice subsequent to deposition (Figure 67).

In rare cases, the ice-contact deposits may be mixed with other deposits that formed simultaneously. Such a deposit is illustrated in Figure 68, which shows valley-side kame sands mixed

Table 5. *Summary of Composition of Sand-Sized Material in Deposits of Olean Ice-Contact Stratified Sand and Gravel in Pike County*

(Values represent percentage of total composition found on sieve[1])

| Sieve size[3] (number) | Quartz[2] | | | | Nonred sandstone and siltstone[2] | | | | Red sandstone and siltstone[2] | | | |
|---|---|---|---|---|---|---|---|---|---|---|---|---|
| | A | B | C | D | A | B | C | D | A | B | C | D |
| 8 | 1 | 1 | Tr[4] | 1 | 99 | 99 | 78 | 58 | Tr | Tr | 22 | 41 |
| 16 | 1 | 2 | 6 | 10 | 99 | 94 | 70 | 50 | Tr | 4 | 24 | 40 |
| 30 | 7 | 8 | 29 | 48 | 93 | 91 | 49 | 30 | Tr | 1 | 22 | 22 |
| 50 | 50 | 45 | 69 | 77 | 50 | 54 | 15 | 10 | Tr | 1 | 16 | 13 |
| 100 | 71 | 83 | 85 | 88 | 29 | 16 | 6 | 5 | 0 | 1 | 9 | 7 |
| 200 | 73 | 91 | 89 | 92 | 27 | 7 | 4 | 4 | 0 | 2 | 7 | 4 |
| Finer[5] | 70 | 93 | 90 | 93 | 30 | 5 | 3 | 4 | 0 | 2 | 7 | 3 |

[1]Determined by visual estimation.
[2]Lithologies are subdivided by underlying bedrock as follows: A—Trimmers Rock and Mahantango Formations (3 samples); B—Towamensing Member of Catskill Formation (4 samples); C—Lackawaxen and Delaware River Members of Catskill Formation (22 samples); and D—Poplar Gap and Packerton Members, undivided of Catskill Formation (11 samples).
[3]See Table 6 for millimeter size of each sieve number.
[4]Tr, trace.
[5]This category includes all material passing through a number 200 sieve.

Figure 63. Olean ice-contact stratified sand exposed on the west side of Kleinhan Creek about 1.1 miles west of Bear Top Mountain at 41°22′01″N/75°14′57″W.

Figure 64. Rippled Olean ice-contact stratified sand exposed on the northeast bank of Rattlesnake Creek about 0.3 mile north-northwest of Sunrise Lake at 41°19′36″N/74°57′58″W. The scale is divided in 1-foot intervals.

with shale-chip rubble. The shale-chip rubble was apparently accumulating at the base of exposed cliffs along the Delaware River valley at the same time that the valley-side kame was being deposited adjacent to melting ice.

## SHALE-CHIP RUBBLE

INTRODUCTION: Shale-chip rubble is a frequently utilized but generally unheralded economic resource in Pike County and other areas of Pennsylvania. General information on this material can be found in Sevon and Berg (1979). Its resource potential is discussed in the section "Mineral Resources."

FIELD OCCURRENCE: In Pike County, shale-chip rubble occurs mainly on the steep slopes along the northwest side of the Delaware River valley between Bushkill and Matamoras (Crowl,

Figure 65. Interlaminated sand, silt, and clay in a deposit of Olean ice-contact stratified sand. The exposure is about 0.15 mile west of Pa. Route 507 in Greentown at 41°18′24″N/75°18′37″W. The length of the scale is about 25 inches, and the divisions are in inches.

Figure 66. Small normal faults developed as a result of collapse in a deposit of Olean ice-contact stratified sand and gravel exposed on the west side of Blooming Grove Creek about 1 mile northwest of Lake Giles at 41°23′13″N/75°06′40″W. The scale is divided in inches and centimeters.

Figure 67. Folded beds of Olean ice-contact stratified sand and gravel exposed on the east bank of Wallenpaupack Creek about 2.9 miles south-southwest of Greentown at 41°17′03″N/75°19′44″W. The folds indicate override of ice from right to left, approximately north to south. The pick handle is about 16 inches long.

Figure 68. Interbedded shale-chip rubble and Olean ice-contact stratified sand exposed in a borrow area along the west side of U.S. Route 209 about 1.6 miles southwest of Milford at 41°18′10″N/74°49′18″W. The hammer is 11.5 inches long.

1971). Some shale-chip rubble also occurs on similar slopes along the northwest side of Bushkill Creek southwest of Bushkill. The slopes are formed by the claystone and siltstone of the Mahantango Formation. In undisturbed settings, the shale-chip rubble covers most of the bedrock slope except for some near-vertical free faces near the tops of the slopes.

DESCRIPTION: Shale-chip rubble is made up of angular, elongate to platy fragments of claystone and siltstone generally 2 to 5 cm long (Figure 69). These fragments are loosely packed and generally have no matrix material between them, but most have a thin coating of clay. In cross section normal to the bedrock slope, the fragments are arranged in poorly to well-defined layers (Figure 70) and generally are well sorted within a specific layer. The layers range in thickness from less than an inch for a single fragment to 4 inches. Platy fragments in these layers are in many places imbricated, having an upslope dip. Elongate fragments generally have their long axis parallel to the slope. The layers of fragments dip steeply away from the rock wall (Figure 70), and slope angles greater than 30 degrees are not uncommon. A few layers have an abundance of clay matrix, the interstitial space between fragments being nearly or completely filled, but such layers are not common.

Size sorting within individual layers is good to excellent, but large fragments, up to several tens of centimeters long, occur sporadically throughout the deposits. Variations in fragment size between well-defined layers is common and may be large. Some layers have an upward coarsening of fragment size.

The deposits of shale-chip rubble are basically wedge shaped, having the thickest part of the wedge at the base of the slope and the thinnest part at the top of the slope. The upper surface of the deposits is well vegetated (Figure 71) and totally masks the nature of the deposits except where excavation has occurred. In some places, accumulation of rubble is occurring on the vegetative cover below exposed bedrock ledges (Figure 72).

THICKNESS: The thickness of deposits of shale-chip rubble is variable, both within a specific deposit and between deposits. It may exceed 50 feet at the base of the slope in some deposits. The thickness decreases upslope to the top of the deposit wedge. The deposits also thin toward their lateral margins. The uniformity of the upslope change in thickness is dependent upon the shape of the underlying bedrock slope. In some cases the bedrock has a uniform steep slope, and the rubble thickness decreases uniformly. In other cases the bedrock has a stair-stepped character, and the thickness of the overlying rubble is variable. The length of a slope covered by rubble commonly exceeds 200 feet.

CONTACTS: The steep upslope contact of the rubble deposits is with the underlying Mahantango Formation. This contact is sharp and conforms to the configuration of the underlying bedrock. In some fresh exposures where rubble deposits have been removed, the bedrock is polished and striated. The valley-bottom contact of the rubble wedge is generally covered, and the underlying material is unknown. Locally the rub-

Figure 69. Breakup of Mahantango Formation claystone into elongate fragments. The fragments move downslope and accumulate as shale-chip rubble. The bedding dips into the rock face; the cleavage is parallel to the rock face. The exposure is in an old borrow area on the west side of U.S. Route 209 about 2.7 miles southwest of Dingmans Ferry at 41°10′54″N/74°53′33″W. The scale is divided in inches and centimeters.

ble is interbedded with the glaciofluvial and glaciolacustrine deposits of the Delaware Valley (Figure 68). The upper surface of the rubble deposits is covered only by vegetation.

EXPOSURES: Exposures of shale-chip rubble are common in numerous borrow areas along the northwest side of U.S. Route 209 between Bushkill and Matamoras. The quality of the exposures is variable and changes frequently because of the susceptibility of the disturbed rubble to slope failure.

ORIGIN: The deposits of shale-chip rubble along the Delaware River valley result from the accumulation of shale fragments on bedrock slopes left bare by the melting of the Woodfordian ice sheet. During the Woodfordian ice advance, the valley walls were eroded, steepened, polished, and striated. As soon as the ice melted sufficiently for the Mahantango shale to be exposed, the rigorous periglacial climate caused rapid breakup of the shale along the numerous bedding, cleavage, and fracture planes present in the rock (Figure 73). This mechanical disintegration, primarily by freeze-thaw action, produced fragments of shale which fell or slid downslope (Figure 69). These slope deposits were either interbedded with whatever material was being deposited in the valley bottom or they accumulated on top of already deposited materials. As the rubble deposit grew in size, the upslope margin of the deposit migrated upward, covering more

Figure 70. Cross section of a deposit of shale-chip rubble exposed in the old borrow area shown in Figure 69. The scale is divided in inches and centimeters.

and more bedrock and thus eliminating part of its source. As the climate became milder following deglaciation, the rate of mechanical disintegration decreased and less rubble accumulated. Vegetation became established and gradually covered and stabilized the slopes.

In borrow areas where the shale-chip rubble has been removed and bedrock is exposed, new deposits of rubble are accumulating at the base of the slope (Figure 71), the result of present-day mechanical disintegration of the shale. The undercutting action of modern quarrying causes additional material to be contributed from shale-chip rubble exposed high on the slope.

## BOULDER ACCUMULATIONS

INTRODUCTION: West of Pike County, Sevon (1975a, b) mapped some boulder deposits as boulder fields and boulder colluvium. These deposits were differentiated as follows (Sevon, 1975a):

> Boulder field is differentiated from boulder colluvium because (1) a boulder field is a thick deposit constituting an accumulated mass of boulders, (2) boulder fields have no interstitial material at the surface; interstitial material at depth and adjacent boulders, even when surrounded by interstitial material, are in point contact, (3) boulder fields are barren except along the margins where a vegetative mat overlies voids and plants have been established, (4) boulders in boulder fields often have polished and "fitted" surfaces indicative of grinding of adjacent boulders, and (5) boulder fields generally occur on lower slope angles than boulder colluvium. Boulder fields represent boulder accumulations in valley bottoms as a result of periglacial action . . . .
>
> Boulder colluvium is differentiated from boulder field because (1) boulder colluvium is presumably a thin deposit comprising only a surface cover, (2) boulder colluvium often has abundant interstitial material at the surface as well as below the surface and adjacent boulders may or may not be in point contact, (3) boulder colluvium supports a variety of vegetation, (4) boulders in boulder colluvium are not "fitted," ground, or polished, and (5) boul-

Figure 71. Mahantango Formation rock and shale-chip rubble exposed in a borrow area on the west side of U.S. Route 209 about 5.1 miles southwest of Dingmans Ferry at 41°09′30″N/74°54′43″W. The vegetated upper surface is on an old, undisturbed deposit of shale-chip rubble; the shale-chip rubble at the base of the rock face is accumulating at present.

> der colluvium generally occurs on steeper slopes than boulder fields. Boulder colluvium represents material derived from outcrops on moderate to steep slopes, moved downslope by periglacial gravity action without accumulating an appreciable thickness of boulders. Boulder colluvium commonly grades into boulder fields.

This usage was followed in mapping of nearby areas by Berg (1975), Berg and others (1977), and Sevon and Berg (1978). Sevon (1975c) combined the two units into boulder field-colluvium as follows:

> The material here called boulder field-colluvium combines two separate units, boulder fields and boulder colluvium, mapped by the writer to the southwest in the Hickory Run area (Sevon, 1975b). In that area the two deposits were differentiated mainly on the basis of differences in thickness, slope position, and vegetation. In this report area, differences in thickness and slope position are minimal and generally do not warrant differentiation. Most of the boulder field-colluvium is vegetated and only intermittent areas are completely free of vegetation, a criterion previously required for designation as a boulder field (Sevon, 1975b).

In addition, Sevon (1972) identified a positive colluvial origin for boulder accumulations developed on and from till of Woodfordian age. Stone stripes and polygons developed on the surface of till are the evidence for a colluvial origin.

However, examination of numerous concentrations of boulders in Pike County indicates that many of these concentrations were formed by glaciation rather than colluviation. Therefore, the authors reject the usage of the term "colluvium" for these deposits and use instead the nongenetic term "accumulation." Thus, they would substitute the name boulder accumulation for both boulder colluvium and boulder field-

Figure 72. Accumulation of shale-chip rubble occurring on vegetated slopes below a ledge composed of rock of the Mahantango Formation. The exposure is on the west side of Interstate Route 84, opposite the Delaware Valley High School, southwest of Matamoras at 41°21′10″N/74°44′22″W.

colluvium and retain the term boulder field as defined by Sevon (1975a).

Boulder accumulations are common throughout Pike County, but only some of the larger identified accumulations have been mapped (Plate 2). Many deposits are too small to show at the map scale (1:50,000), and many cannot be readily identified on aerial photographs.

FIELD OCCURRENCE: Boulder accumulations occur in topographically low areas, such as small drainageways (Figure 74), in relatively flat areas (Figure 75), and below rock outcrops (Figure 76). No differentiation of the occurrences has been made on Plate 2. The accumulations occur mainly in areas underlain by the Catskill Formation (Area A in Figure 50). They are commonly present around the margins of peat deposits and may underlie some peat deposits. Boulders are also common on swamp surfaces and may represent accumulations partly buried by the swamp deposits.

Throughout the county, many steep slopes with few if any rock outcrops have a surface distribution of boulders that varies from a sparse to a nearly complete covering. Such deposits have not been mapped, but many are of sufficient areal size and boulder concentration to be considered boulder accumulations.

DESCRIPTION: Boulder accumulations contain an unsorted mixture of angular to rounded boulders of varying size (Figure 77) concentrated on the surface in areas of varying size and shape. Some boulders are as large as 6 feet or more in diameter, but most are less than 3 feet. The boul-

Figure 73. Mahantango Formation showing prominent parting surfaces of bedding (B), fractures (F), and spaced cleavage (C), along which the shale disintegrates into rubble fragments (lower left). The scale is divided in inches and centimeters. The bedding strikes N50°E and dips into the rock face at 13°N; the spaced cleavage strikes N70°E and dips 68°S; the fractures strike N30°W and dip 72°SW. The outcrop is located along a steep, private access road in a large abandoned borrow area immediately southwest of Matamoras at 41°21′29″N/74°43′15″W.

ders are composed dominantly of sandstone and vary from blocky to platy in shape. Some boulders have very angular edges, but most have slightly rounded edges and some have well-rounded edges. Some accumulations have a restricted range of boulder size and appear relatively uniform, whereas other accumulations have a large range of boulder size and surfaces that appear very rough and irregular. Most deposits have a moderate to abundant amount of sandy soil and organic debris between boulders below the surface, but the surface layer of boulders frequently lacks matrix. The boulders in boulder accumulations in Pike County appear to be randomly distributed with regard to size and long-axis orientation, but no studies have been performed to verify this field observation.

The accumulations that occur in the low part of small drainageways are generally relatively flat deposits (Figure 74), although some accumulations are higher in the center than along the margins (Figure 78). These accumulations have a moderate range of boulder sizes and generally lack very large boulders. In some places, they are traversed by a well-defined stream course, but in most places there is no evidence of movement of the boulders by water. Some boulder accumulations grade into less closely packed accumulations on adjacent slopes, but generally boulders are sparse on these slopes. The accumulations are

Figure 74. Accumulation of sandstone boulders occurring in a topographic low between adjacent slopes about 2.3 miles south-southeast of Twelvemile Pond at 41°08′42″N/75°04′15″W. The scale is divided in feet.

Figure 75. Accumulation of platy sandstone boulders occurring on a flat area about 1.7 miles southwest of McCartys Corner at 41°16′48″N/74°53′28″W. The scale is divided in feet.

commonly well vegetated by trees and some understory of bushes, grasses, and ferns.

The accumulations that occur in relatively flat areas are generally composed of boulders of relatively uniform size and have the flattest surfaces (Figure 75). These accumulations are gradational into less bouldery marginal areas. They frequently cover small- (few feet) to moderate-sized areas (few tens of feet) that are free of vegetation, and the lateral decrease in the quantity of boulders into marginal areas is accompanied by an increase in vegetative cover.

The accumulations that are downslope from rock outcrops have the greatest range of boulder size and the poorest boundary definition (Figures 48 and 76). The largest boulders are commonly close to a rock outcrop, and generally boulder size decreases with increasing distance from the

Figure 76. Accumulation of sandstone and conglomerate boulders occurring on a shallow slope south of a sandstone and conglomerate outcrop at the north end of White Deer Lake at 41°24′39″N/75°07′49″W. The scale is divided in feet.

outcrop. These deposits are less closely packed than the deposits in small drainageways and on flat areas, and soil-organic matrix at the surface is common. Because of the diversity of sizes present, these accumulations resemble a randomly jumbled mass of rocks lacking the flat surface typical of the other two types.

THICKNESS: The thickness of the boulder accumulations is unknown. Some accumulations appear to be little more than a surface coating of boulders, whereas some accumulations in steep-gradient and steep-walled drainageways appear to be several feet thick. No artificial cuts through any of the accumulations were observed during mapping.

CONTACTS: The basal contact of a boulder accumulation has not been observed, but all accumulations are presumed to rest on either glacial till or bedrock, and the contact is presumably irregular. The accumulations are overlain only by peat or swamp deposits in some places.

EXPOSURES: Numerous good exposures are easily accessible throughout the county and in-

Figure 77. Rounded, angular, and platy boulders in a boulder accumulation at the north end of White Deer Lake at 41°24′39″N/75°07′49″W. The scale is divided in feet.

clude the locations given in Figures 74 to 78. An excellent exposure occurs at the north end of White Deer Lake (Figure 76). Numerous boulder-accumulation deposits occur along the road on the north and east margins of Promised Land Lake in Promised Land State Park.

ORIGIN: Boulder accumulations are thought to be the product of glaciation and/or colluviation. During Woodfordian glaciation, ice erosion of the edges of numerous rock ledges apparently was severe, and much of the eroded material, in the form of boulders of varying size, was deposited near the rock ledges. In addition, topographic depressions and drainageways apparently acted as sediment traps for boulders entrained at the bottom of the glacier. Following deglaciation, some frost riving of exposed rock and some downslope mass movement of rock debris occurred. The degree of importance of each of the suggested mechanisms of origin is not known, but the authors think that glaciation is the primary mechanism.

## ALLUVIUM AND OLEAN OUTWASH, UNDIFFERENTIATED

FIELD OCCURRENCE: Olean outwash consists of sand and gravel deposited by streams flowing from the melting Wisconsinan glacier and is found as valley-bottom fill in the Delaware River valley, particularly between Matamoras and Bushkill, and in other smaller valleys in Pike

Figure 78. Part of an accumulation of sandstone boulders occurring in a topographic low between adjacent slopes. The accumulation is mounded in the center and has intermittent streams along both sides. The stream course along the east side occurs just to the left of the scale, which is divided in feet. The accumulation is 2.4 miles south-southeast of Twelvemile Pond at 41°08′47″N/75°04′28″W.

County. The outwash sand and gravel are overlain by alluvial sand deposited by floodwaters. Outwash in the Delaware River valley underlies a terrace about 25 feet above low-water level and narrow floodplain areas 5 to 15 feet above low-water level.

DESCRIPTION: Olean outwash is composed of stratified material ranging in size from very fine grained sand to cobbles and rarely boulders. The sand, gravel, and cobbles are generally mixed together, and the sand content ranges from 30 to 70 percent. The gravels and cobbles are moderately well to well rounded. Stratification, both planar and cross, is generally well developed, and the average clast size may vary considerably from bed to bed. Lithology analyses performed by Crowl (1971) of pebbles 2.5 to 5 cm in size in three samples from the Delaware River valley indicate the following derivation of clasts: 71 percent, Catskill Formation; 17 percent, Shawangunk Formation; 4 percent, Mahantango Formation; and 8 percent, miscellaneous. Larger sizes are mainly derived from Catskill sandstone and Shawangunk sandstone and conglomerate. Fragments of Mahantango siltstone are more abundant in sizes less than 2.5 cm and persist into the coarser sand sizes. These deposits are generally unconsolidated, but some calcite cementation occurs locally where moderate quantities of limestone clasts are present.

Little is known about the textural variations of outwash below the surface. Drilling information in the Delaware River valley at the proposed Tocks Island Dam site southwest of Pike County (U.S. Army Corps of Engineers, 1967; Crowl,

1971) indicates considerable textural variation and the presence of silts and clays of presumed glaciolacustrine origin. No information is available elsewhere.

Alluvium comprises mainly very fine to fine-grained sand, some silt, and minor amounts of clay. It is generally reddish brown to brown, shows only a slight variation in texture, and has minimal stratification. The sand is friable and well drained and corresponds to soils of the Tioga series mapped in Pike County (Taylor, 1969). Pebbles and cobbles occur within the sands, but are not abundant.

THICKNESS: The maximum known thickness of alluvium and Olean outwash in the Delaware River valley is 160 feet in a well at Milford. Thickness values from other wells along the Delaware River valley are as follows (Lohman, 1937; Davis, 1989): Lackawaxen, 75 feet; Shohola, 135 feet; Millrift, 100 feet; Matamoras, 27 to 84 feet; Dingmans Ferry, 42 feet and 130 feet; and near Bushkill, 20 feet and 108 feet. Two wells in the lower Lackawaxen River valley have an average thickness of 168 feet (Davis, 1989).

The thickness of alluvium and Olean outwash in other parts of Pike County is not known, but is presumably in the range of the above values. In all of the measured deposits, the bulk of the material is outwash, and alluvium constitutes only a thin veneer on the top.

CONTACTS: The basal contact of deposits of alluvium and Olean outwash is either with glacial till or bedrock. The deposits are not overlain by any other material.

EXPOSURES: Exposures of alluvium and Olean outwash are not abundant and are generally temporary. Cutbanks along streams may expose some materials, but the amount exposed is small.

ORIGIN: Olean outwash was deposited by streams flowing from melting glacial ice. The outwash streams sometimes covered stagnant blocks of ice, which later melted to form small ponds in which finer grained sediments (glaciolacustrine) were deposited. Subsequent to deglaciation in the Delaware River valley, erosion of some of the outwash created an outwash terrace. Since deglaciation, repeated floods have deposited alluvial sand on top of the coarser outwash materials on the floodplain. Streams are presently eroding the valley-fill materials.

MISCELLANEOUS: Olean outwash is an extensive source of sand and gravel and is also a source of groundwater.

## ALLUVIUM AND ALLUVIAL FANS

FIELD OCCURRENCE: Alluvium occurs in the bottoms of many valleys throughout Pike County. Alluvial fans occur at the mouths of some streams where they open into larger valleys. In some places, the alluvium may overlie glacial outwash, but the actual existence of outwash was not verified because of the lack of exposure. Alluvium was mapped only where feasible at the map scale (1:50,000) (Plate 2).

DESCRIPTION: Alluvium comprises an unconsolidated, poorly stratified, poorly to moderately well sorted mixture of cobbles, gravel, sand, silt, and clay. Boulders are present in some deposits. The larger clasts range from well rounded to angular and platy. The variation in grain size between beds ranges from minimal to marked. Elongate and platy clasts are commonly imbricated and have an upstream dip. Some deposits have a thin (3 feet or less thick) layer of sand and silt in the uppermost part, but much of the alluvium is composed of a mixture of sizes. The composition of an alluvium deposit reflects the lithologies present in the local bedrock. This is true even though much of the alluvium is eroded glacial material (which also strongly reflects the underlying bedrock). Both the texture and lithologic composition of deposits of alluvial fans is the same as alluvium.

The surface of alluvium deposits varies from relatively flat and smooth to very rough. Flat-surfaced deposits of alluvium occur mainly in larger drainages having moderate to low gradients, moderate widths, and perennial streams. Small, steep-gradient valley bottoms, commonly having intermittent streams, generally contain rough-surfaced deposits of alluvium made up of many small lobes with steep front and side margins and rounded to flat tops. In some deposits, the lobes terminate on the downstream end at a tree and on the upstream end at a small depression. Smooth-surfaced deposits of alluvium are generally well covered with vegetation ranging from grass to trees. Rough-surfaced deposits of alluvium normally have less vegetation, most of which is small shrubs and trees.

Alluvial fans occur at the mouths of streams where a smaller valley joins a larger valley. The surface of an alluvial fan is relatively smooth, has a fanlike shape, and slopes away from the center of the smaller valley. The gradient of the fan surface is dependent mainly on the gradient of the smaller valley: the steeper the gradient of the source stream, the steeper the gradient of the alluvial fan. Most alluvial fans in Pike County are vegetated and are undergoing stream erosion.

THICKNESS: The thickness of alluvium in Pike County is generally unknown because exposures are relatively small and do not show all of a deposit. Estimates suggest that alluvium is not more than a few feet thick in most places. Valleys that have undetected glacial deposits beneath alluvium cover would have thicker amounts of unconsolidated material in the valley bottoms.

CONTACTS: The basal contact of alluvium is either with bedrock or underlying glacial deposits, either outwash or till. The alluvium deposits are not overlain by any other material.

EXPOSURES: Most of the streams in Pike County are presently eroding their channels, and small exposures generally less than 3 feet high occur along the banks of some streams. However, because of slumping along the banks and growth of vegetation, these exposures are temporary.

ORIGIN: Alluvium has been formed by fluvial deposition in valley bottoms since deglaciation occurred in Pike County at least 18,500 years ago. Much of the alluvium was probably deposited soon after deglaciation, before the landscape was covered with vegetation. Subsequent growth of extensive plant cover reduced erosion until the land was cleared during the nineteenth century.

The layer of fine-grained sediments that overlies coarser sands and gravels in some alluvium deposits represents material deposited in the overbank area during floods. The small lobelike features found in steeper gradient, small valleys formed during periods of large-volume water flow by either debris flow from an immediately upstream position or incremental accumulation of material behind an obstruction.

## SWAMP DEPOSITS

INTRODUCTION: Swamp deposits are variable in surface expression and physical makeup, but are a mappable geomorphic feature which occurs throughout Pike County. The unit corresponds approximately to soils of the Norwich and Tughill series mapped in the county (Taylor, 1969, p. 58, 65–66).

FIELD OCCURRENCE: Swamp deposits occur in relatively flat, undrained or poorly drained areas throughout Pike County, but are most common in areas underlain by rocks of the Catskill Formation. The deposits occur in several settings: in isolated, low-relief depressions, along the margins of peat deposits, and on valley floors adjacent to through-flowing streams. Most of the mapped occurrences were not visited in the field, but were located by interpretation of color aerial photographs and were differentiated from peat deposits on the basis of size and density of tree cover. Some areas mapped as swamp deposits may be peat deposits and vice versa (Plate 2).

DESCRIPTION: Swamp deposits comprise unconsolidated clay, silt, and sand mixed with muck, a finely divided, partly decomposed organic material derived from plant debris. Pebbles and cobbles may occur within the deposit, and some thin deposits are very bouldery and have numerous moss-covered boulders on the surface (Figure 79). Deposits on valley floors generally have few or no boulders on the surface (Figure 80). Material in swamp deposits is very dark gray (N3) to black (N2.5/0) at the surface and dark grayish brown (2.5Y4/2) at depths of

Figure 79. Surface of a swamp deposit having abundant moss-covered boulders and dead, fallen trees. The area is usually covered with water except during dry periods of autumn. The swamp is located about 1.4 miles west of Germantown at 41°25′04″N/75°06′00″W. The scale is divided in feet.

20 inches or more. The material is generally sticky and variable in coherence depending on the amount of clay present. Some deposits have a thin veneer of peat and/or muck at the surface. Some sites have shallow, stagnant water over most of the area at all times except during extended dry periods, whereas other sites have little or no water standing on the surface, but are water saturated immediately below the surface. Deposits containing surface boulder litter tend to have standing water, abundant sphagnum moss, a minimal understory of shrubs, and some fallen trees (Figure 79). Deposits lacking surface boulder litter and standing water tend to have a variety of low, water-loving plants, such as skunk cabbage, and a minimal understory (Figure 80). All of the deposits have a thick overstory of trees, particularly conifers.

THICKNESS: Most swamp deposits in Pike County are thought to be less than 6 feet thick. No attempt was made to probe the deposits to determine thickness.

CONTACTS: The basal contact of swamp deposits is presumably sharp, but has not been observed. The contact is presumably with either bedrock or glacial drift.

EXPOSURES: Swamps occur throughout Pike County, but examination of the deposit materials requires excavation, augering, or coring. Readily accessible sites are indicated in Figures 79 and 80.

Figure 80. Surface of a swamp deposit having skunk cabbage and other wet-area plants. Note the relative absence of boulders. The surface is usually damp, and has standing water only during the wetter parts of the year. The swamp is about 2.5 miles south of Twelvemile Pond at 41°08′51″N/75°04′12″W. The scale is divided in inches and 0.5-foot intervals.

ORIGIN: Swamp deposits originate from the mixing of materials washed into undrained or poorly drained areas from higher adjacent areas with the partially decayed organic debris that accumulates in the wet environment.

## PEAT

INTRODUCTION: Peat deposits occur in poorly drained glaciated depressions throughout most of Pike County. Mapping of peat deposits was done mainly by interpretation of aerial photographs and some supplementary field examination. The boundaries of mapped peat deposits were placed at marked changes in vegetation observed on color aerial photographs and may not everywhere coincide with the actual subsurface distribution of peat accumulations.

FIELD OCCURRENCE: Cameron (1970, p. A26–A37) recognized five types of depressions created by glaciation where peat accumulates: (1) a bedrock trough closed at both ends by glacial drift; (2) a bedrock trough closed at both ends and partly mantled by glacial drift; (3) a depression in bedrock closed by glacial drift; (4) a depression in glacial drift closed by bedrock; and (5) a depression in glacial drift. All of these types of depressions occur in Pike County, but the mapped peat deposits (Plate 2) have not been categorized according to the types. Examples of the types are as follows: (1) Smiths Swamp, 2 miles southeast of Germantown at 41°23′10″N/75°03′43″W (approximate center of bog); (2) Big Swamp, 3.8 miles south of Porters Lake at 41°11′43″N/75°05′42″W; (3) Big Inlet Swamp, 3.2 miles east of Promised Land at 41°18′54″N/75°09′00″W; (4) Little York Swamp, 2.5 miles south of Lords Valley at 41°19′54″N/75°04′33″W; and (5) Nebo Swamp, 2.9 miles southeast of Porters Lake at 41°12′50″N/75°02′07″W. Figure 81 is a stereogram showing several different peat deposits, and Figure 82 is a diagrammatic summary of a typical occurrence of peat.

DESCRIPTION: Peat is made up of partly decomposed mosses, ferns, grasses, rushes, reeds, sedges, shrubs, and trees. Three types of peat occur in Pike County: moss peat, reed-sedge peat, and peat humus. Limited information (Edgerton, 1969; Cameron, 1970) indicates that reed-sedge peat is the most abundant type. Moss peat includes mainly sphagnum, hypnum, and other mosses. Sphagnum-moss peat is light tan to brown, light in weight, and porous. It has a high water-holding capacity, high acidity, and a low nitrogen content. Reed-sedge peat is very fibrous and brown to reddish brown. It has a medium water-holding capacity and a medium nitrogen content. Peat humus is in such a sufficiently advanced stage of decomposition that original plant remains are not identifiable. It is dark brown to black, has a low water-holding capacity, and has a medium to high nitrogen content.

Analyses of Pike County peat by Cameron (1970) indicated an average organic content of 78.5 percent (range of 35 to 99 percent), an average water-holding capacity of 1,692 percent (range of 570 to 2,500 percent), an average fiber content (organic particles longer than 0.15 mm) of 45 percent (range of 9 to 71 percent), and an average pH of 5.65 (range of 4.3 to 6.45). Analy-

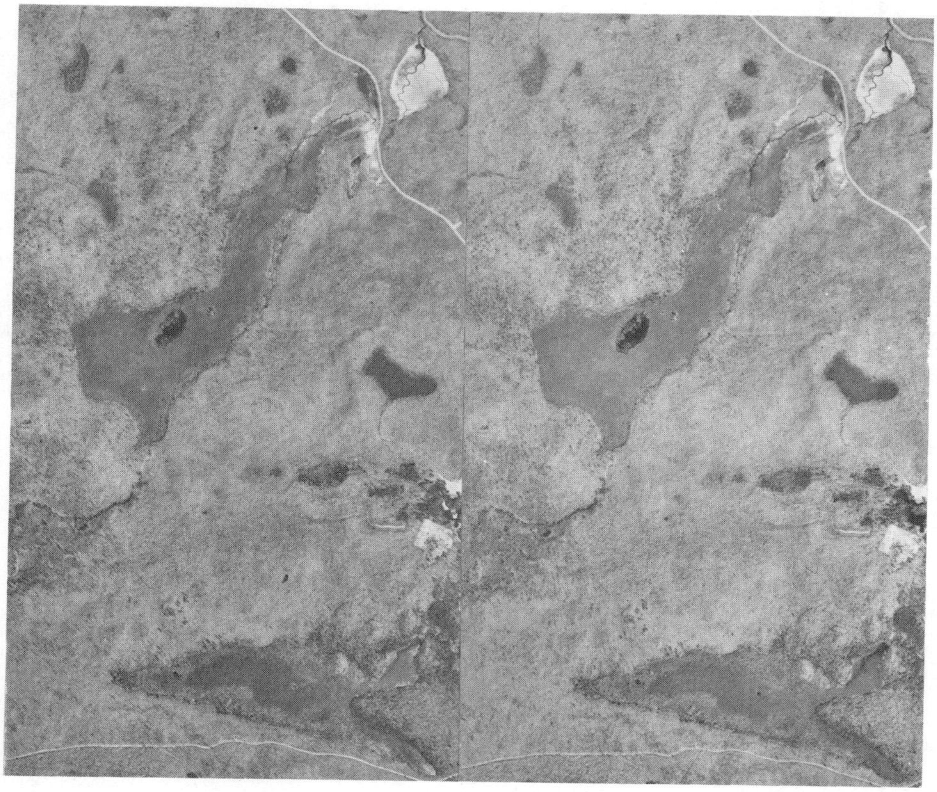

Figure 81. Stereogram of several peat deposits southeast of Germantown (the approximate center of the stereogram is at 41°23′07″N/75°02′50″W). The aerial photographs were taken for the U.S. Geological Survey on April 15, 1973; the scale is approximately 1:20,000. The large peat bog on the left is Little Mud Pond Swamp (see Figure 84), and the large peat bog near the bottom is Smiths Swamp. North is to the left.

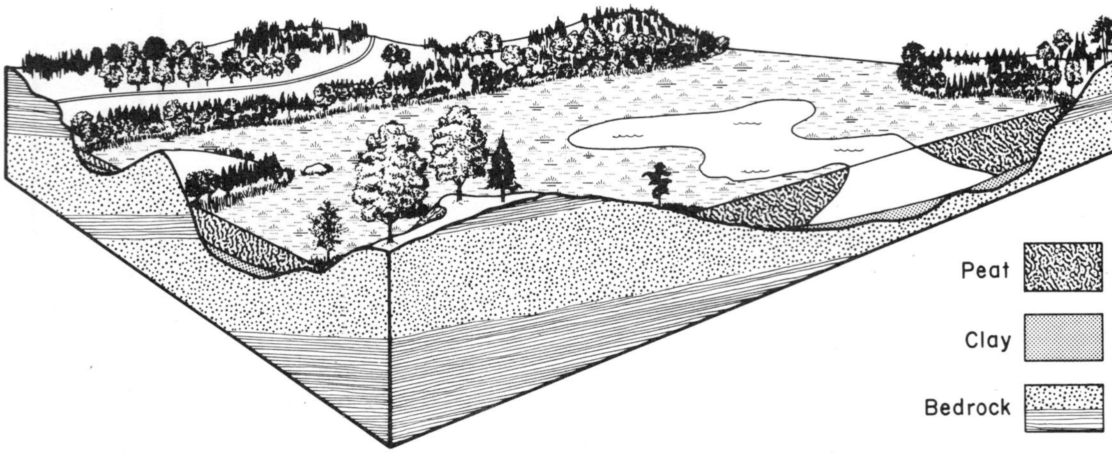


Figure 82. Typical occurrence of peat.

ses of 40 peat samples from Little Mud Pond Swamp (Figure 81) by Palmer and MacNamara (1972, p. 119) yielded an average organic content of 95.6 percent (range of 91.4 to 98.6 percent) for that bog.

All peat deposits in Pike County, except where artificially drained, are water saturated and have the water table at or near the surface. As a result, the peat is very soft and yields readily to any probing or excavating. Drying the peat in place by artificially lowering the water table increases the resistance of the peat to probing and excavation.

The basal configuration of the peat deposits is unknown, but detailed probing of several peat deposits by Edgerton (1969) and Palmer and MacNamara (1972) indicated that the bottom configuration of most deposits is probably irregular. Available data indicate that most, if not all, peat deposits have a layer of clay, sandy clay, or till beneath them. Boulders may underlie the clay throughout the entire area of some peat deposits. Data are not available regarding the thickness of the basal sediments, but they probably are not more than a foot or so thick in most places.

The upper part of many peat bogs in Pike County is black muck, which is very decomposed organic matter lacking fibrous material and having some granular soil structure. Peat deposits generally have some pieces of wood in the upper part (e.g., Figure 84).

The upper surface of peat deposits is flat and level, generally swampy, and may have a small body of open water. The margins of peat deposits commonly have numerous boulders and are laterally transitional into swamp deposits or muck. The surfaces of some peat deposits are relatively open, having a low overgrowth of grass, sedge, moss, and some sparse shrubs (Figure 83); some are partly or entirely covered with low shrubs (Figure 84), principally high-bush blueberry; and some are partly or entirely overgrown with red maple, black spruce, tamarack, alder, hemlock, birch, white pine, rhododendron, and viburnum. This vegetation contrasts with that of adjacent uplands, which are dominated by hard maple and oak. Rhododendron and high-bush blueberry frequently form a dense fringe at the margins of peat bogs. A rising water table has caused the death of trees in some heavily forested bogs, thus creating a swampy landscape such as the one shown in Figure 85. A drop of the water table may allow the reestablishment of a forest, and fluctuations in the water table may lead to the development of cyclic sequences in the peat accumulations. A vertical sequence through any peat deposit generally shows variations in constituent materials; a typical profile is shown in Figure 86.

Figure 83. Peat bog having a surface cover of low grasses, sedge, moss, and local areas of shrubs. The center of the bog is about 3.6 miles southeast of Germantown at 41°22′50″N/75°01′30″W.

Figure 84. Partly excavated peat, surface grasses, and low shrubs on Little Mud Pond Swamp. The photograph site is on the south side of the swamp, about 3.4 miles southeast of Germantown at 41°23′03″N/75°01′43″W.

Figure 85. Dead trees on the surface of a formerly heavily forested peat bog. The photograph site is about 2.1 miles west of Germantown at 41°25′15″N/75°06′45″W.

Although most peat deposits in Pike County are grounded on the bottom, some are at least partly floating on water. Such is the case for the northern part of Little Mud Pond Swamp in the area around the pond. In general, those peat deposits containing bodies of open water are floating in part, and those lacking open bodies of water are partly or totally grounded.

THICKNESS: The thickness of peat deposits in Pike County is variable. All deposits are generally thickest in the center and thinnest at the margins, but irregularities in the configuration of the depression (Edgerton, 1969; Palmer and MacNamara, 1972) result in numerous exceptions. Many of the small peat deposits are probably relatively thin, 3 to 6 feet thick, but those that are areally large are generally more than 10 feet thick. The thickest known peat deposit in Pike County is about 40 feet thick and occurs in the northern part of Little Mud Pond Swamp (Palmer and MacNamara, 1972). Edgerton (1969) and Cameron (1970) indicated a range in thickness for most peat deposits from 5 to 16 feet.

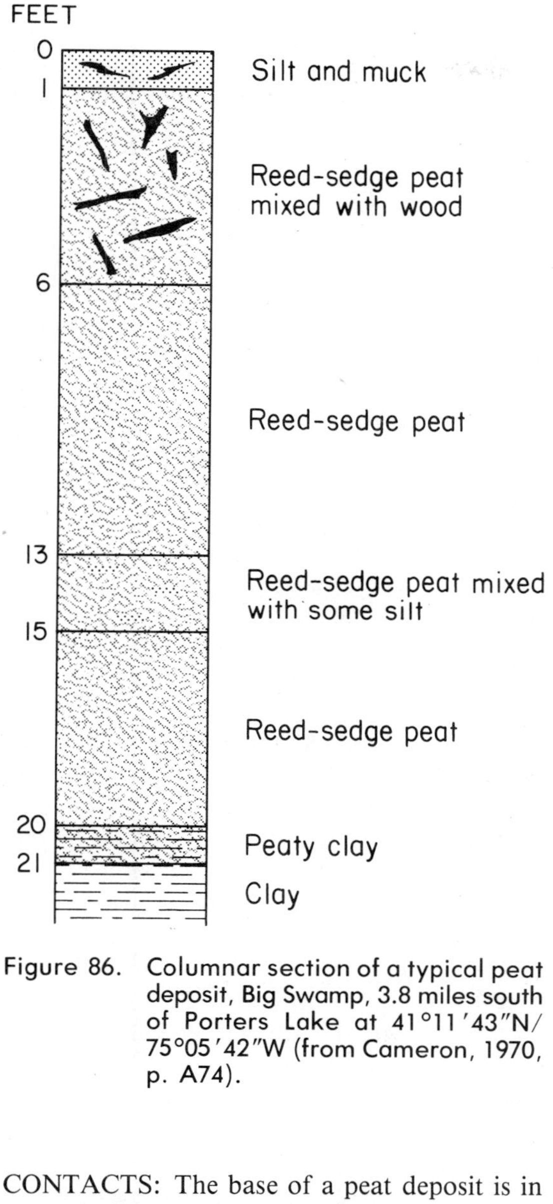


Figure 86. Columnar section of a typical peat deposit, Big Swamp, 3.8 miles south of Porters Lake at 41°11′43″N/75°05′42″W (from Cameron, 1970, p. A74).

CONTACTS: The base of a peat deposit is in contact with gray clay, till, or bedrock. In some deposits there is some interbedding or mixing of the peat and clay.

EXPOSURES: Exposures of peat occur only where harvesting of peat is taking place.

ORIGIN: Peat is an accumulation of organic debris within a shallow depression. Many such depressions were created in Pike County by erosional and depositional events of Woodfordian glaciation. Following deglaciation (at least 18,500 years ago), these depressions were filled with water. As vegetation reestablished itself in the area, aquatic plants, algae, and submerged plants began to grow, and organic debris began to accumulate. Semiaquatic plants became established around pond margins, and, as organic debris continued to accumulate and provide a shallower base, the semiaquatic plants migrated farther into the pond. Eventually plants covered the whole surface, and the area became a marsh. Some of these bogs were eventually covered with forest, and some bogs, as a result of water-table fluctuations, have gone through several cycles of pond, marsh, and forest. This general story of the origin of peat is well illustrated by Cameron (1970, p. A20–A21).

MISCELLANEOUS: Peat deposits in Pike County have a good potential as a resource for soil conditioning. This is discussed in detail in the section "Mineral Resources." Peat deposits also have various environmental and engineering characteristics that are itemized on Plate 2.

## ARTIFICIAL FILL

OCCURRENCE: Fill is used across drainages for leveling roadways. It is also used in dams for lakes.

DESCRIPTION: Artificial fill comprises a mixture of angular rock fragments, sand, silt, and clay, or concrete. This material is compacted, in many places has steep slopes, and is commonly vegetated with grass and crown vetch.

THICKNESS: Artificial fill is commonly several feet thick, and in some places exceeds 30 feet in thickness.

EXPOSURES: Artificial fill is abundant along Interstate Route 84.

ORIGIN: Artificial fill is generally derived from borrow pits or roadway excavations and is compacted by heavy equipment.

## DEBRIS-AVALANCHE CHUTES

Debris-avalanche chutes are long trough-shaped scars on steep slopes. The chutes are generally less than 3 feet deep, only a few feet to a few tens of feet wide, and may be up to a few hundred feet long. They occur in unconsolidated surficial material, generally glacial till. A chute may be relatively free of boulder-sized debris, or it may have varying amounts of debris. Bedrock is exposed on the floor of some chutes. Chutes are commonly bordered by low (generally less than 3 feet) ridges of stones and soil. An irregularly shaped mound of rock and soil debris occurs at the lower terminus of a chute. A chute may be totally free of vegetation if it is recent, or it may be moderately to heavily vegetated if it is many years old.

Debris-avalanche chutes occur on moderate to steep slopes in humid, well-vegetated regions. Chutes result from a rapid-flowage form of landslide (Sharpe, 1938, p. 61–62), a debris avalanche, which occurs when large amounts of concentrated rainfall saturate unconsolidated material on a slope. The rapid-flowage landslide may be initiated anywhere on the slope, but the end result is the creation of a long chute from which all vegetation has been removed. Initially, the mound at the base of a chute comprises a chaotic mixture of soil, rocks, and vegetation. The vegetation will decay in time, and only a mound of debris will remain. Then the chute will gradually be revegetated, the bordering ridges and terminating mound will become subdued, and eventually the debris-avalanche chute will be almost undetectable.

Analysis of color aerial photography of Pike County taken in March 1973 revealed 41 readily detectable debris-avalanche chutes in various stages of revegetation. Many more chutes probably exist, but are so thoroughly revegetated that detection could only occur by careful surface examination of the slopes. The identified chutes are shown on Plate 2.

Two debris-avalanche chutes examined in the field were formed in August 1955 (Hack and Goodlett, 1960, p. 55) and occur on the steep, south-facing slope of East Mountain near Goose Pond. One of the two chutes as it appeared in 1982 is shown in Figure 87.

# STRUCTURAL GEOLOGY

## FOLDS

The basic structure in Pike County is that of a homocline in which all of the rocks dip gently to the northwest. As a consequence, rocks become stratigraphically younger from southeast to northwest across the county. This structural situation remains the same beyond Pike County until the axis of the Wyoming-Lackawanna basin is encountered and dips reverse. The relatively flat nature of the rocks is shown in cross section A–A′ on Plate 1.

The low-dipping beds present throughout most of Pike County have a diversity of strike orientations, as is shown by the rose diagrams in Figure 88. The rocks along the southeastern margin of the county are more steeply dipping and have a more consistent pattern of strike orientation than the rocks in the northern and northwestern parts of the county. The strike of beds in the Bushkill area are the most consistent and show the most eastwardly orientation (mean, N55°E) in the county (Figure 88 M). This reflects the structural influence of the Ridge and Valley province, which extends into the area. Strikes farther to the northeast are not as consistent and have more northerly orientations (Figure 88 E, H, I, J, K, and L). The strike of beds in the north and northwestern parts of the county is quite variable and is basically a reflection of the low dips (Figure 88 A, B, C, D, F, and G).

Superimposed upon the homocline are several third-order (wavelengths from several tens of feet to 0.5 mile) and fourth-order (wavelengths of a few tens of feet) folds. Axes of third-order folds are interpreted. Fourth-order folds are exposed in single outcrops and have planar limbs and narrow hinges. This geometry is characteristic of a kink-band style of folding such as that described by Faill (1973). Figure 89 shows one of the best fourth-order folds seen in Pike County. These small folds are probably more numerous than in-

Figure 87. Scar of a debris-avalanche chute formed in August 1955. The surface has been revegetated mainly by ferns. The length of the chute from the camera position to its end is about 650 feet, and it has a drop in elevation of about 160 feet. The chute is on the south-facing slope of East Mountain about 4 miles south of Promised Land at 41°15′51″N/75°12′36″W. The photograph was taken on July 2, 1982.

dicated on Plate 1, but their detection would require better outcrops normal to strike. The axial traces of nine small folds have a mean orientation of N81°E and a standard deviation of 10 degrees. Seven of the folds plunge 2 to 5 degrees toward the southwest and two plunge 2 degrees toward the northeast. The axial traces of four other folds have a mean orientation of N28°E and a standard deviation of 8 degrees. These folds plunge 2 degrees toward the southwest.

## FAULTS

Faults occur in five areas in Pike County: in the vicinity of Mud Pond Mountain south of Promised Land Lake, above the Delaware River northeast of Bushkill near Egypt Mills, along the Delaware River between Lackawaxen and Millrift, to the northeast of Milford, and to the north of Milford. Actual fault zones can be seen only along the Delaware River.

At Mud Pond Mountain, three faults are thought to account for the offset of a block of conglomeratic sandstone that marks the base of the Poplar Gap and Packerton Members of the Catskill Formation (Plate 1). A cross section through this block indicates a maximum offset of about 200 feet. The nature of the faults is not known, but presumably they are nearly vertical. The easternmost of these faults trends north-south and makes a pronounced linear trace on aerial photographs. The position of the other two faults is interpretative. No actual fault zones have been identified in the field.

Trends of low bedrock ridges developed on the Mahantango Formation above the Delaware River valley northeast of Bushkill give the appearance on aerial photographs of intersections and irregularities possibly related to several faults. However, bedding and cleavage measurements throughout the area and examination of outcrops indicate few bedrock irregularities, and only three faults have been mapped (Plate 1). No fault traces were found, and the faults bounding the small wedge shown on Plate 1 were interpreted from bedding-dip reversals, which do not appear to be related to folding.

Fletcher and Woodrow (1970) mapped four small faults along the Delaware River valley between Lackawaxen and Millrift and one fault south of Shohola on the east side of Shohola Creek. They described the faults as follows (p. 22): "The steep limb of some asymmetrical anticlines is broken by a high-angle, south-dipping reverse fault with the steep south limb thrust toward the foldcrest. Vertical displace-

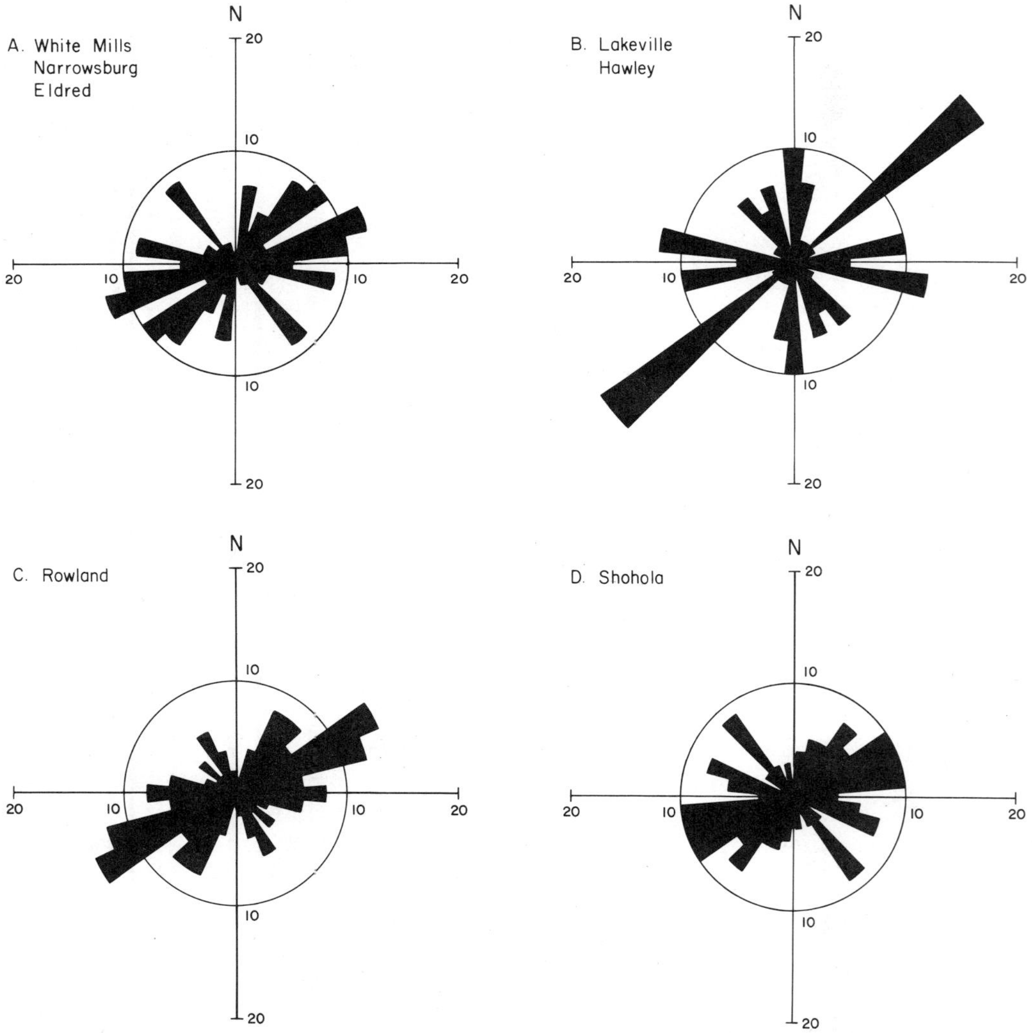


Figure 88. Rose diagrams of bedding-strike orientations in Pike County. Each rose diagram is a summary of measured bedding-strike orientations within the quadrangle or quadrangles indicated.

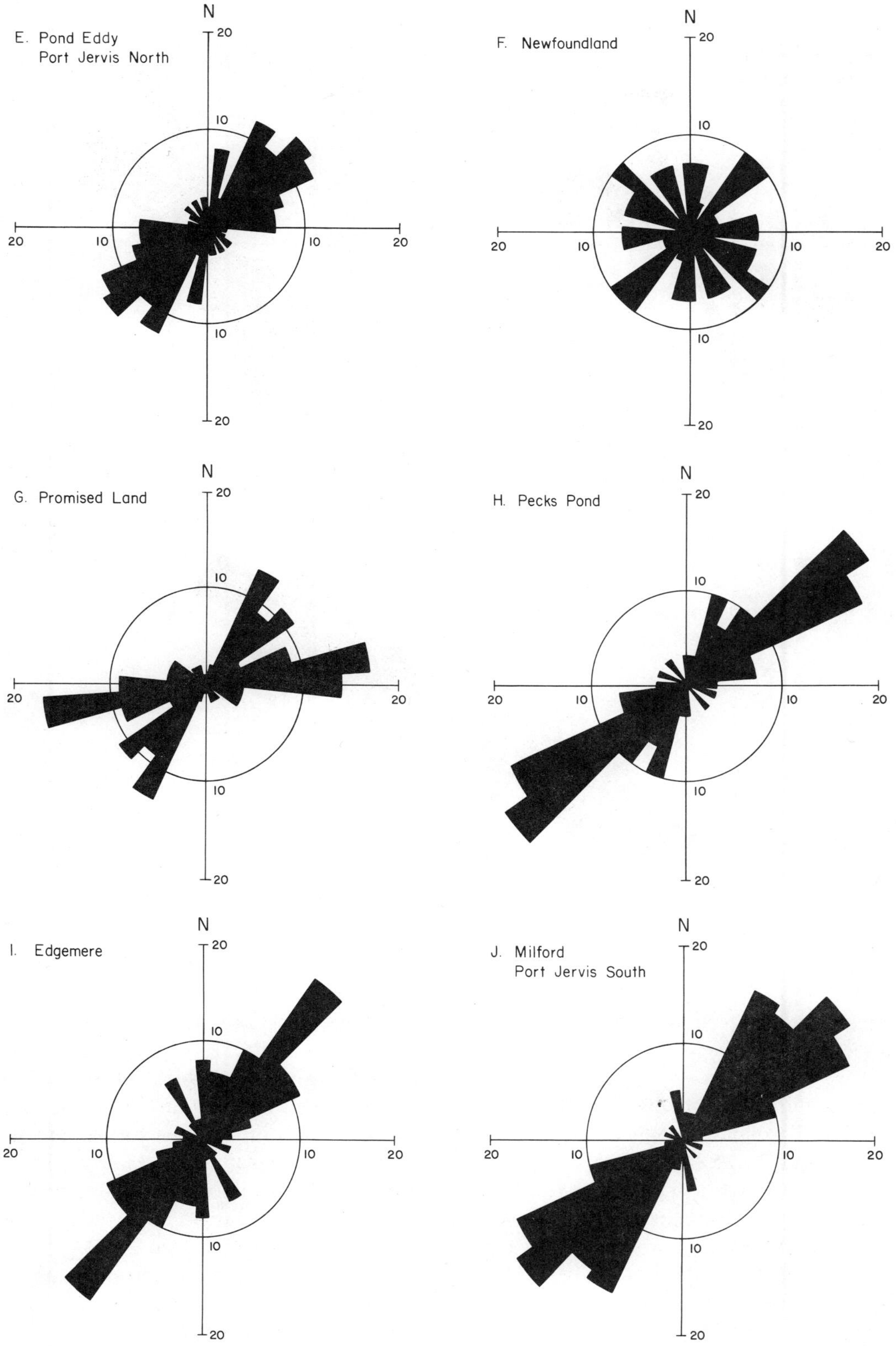


Figure 88. *(Continued).*

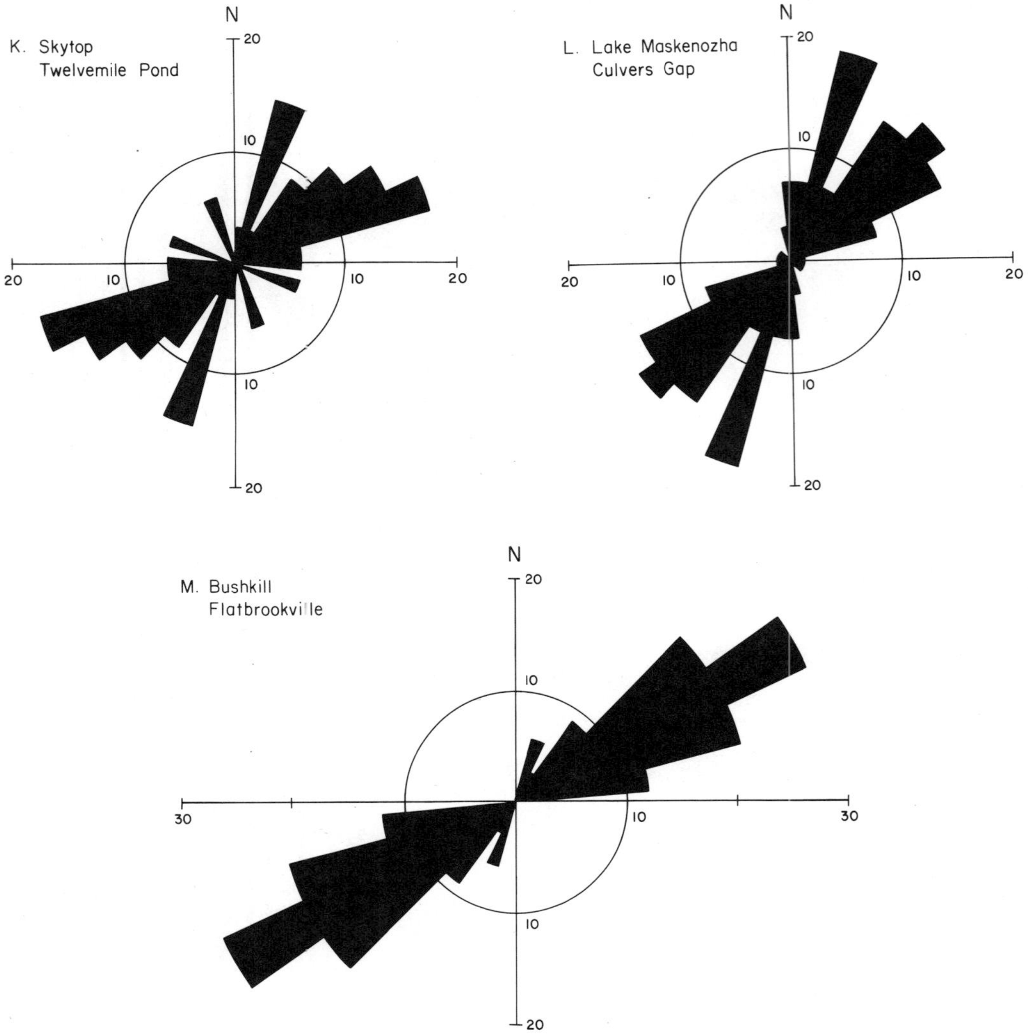


Figure 88. *(Continued).*

ment ranges between 5 and 15 feet." Their mapping is accepted, but the faults are not shown on Plate 1.

A large bedding-plane fault is exposed in the Mahantango Formation along the eastbound lane of Interstate Route 84 northeast of Milford (this is the first long roadcut on both sides of Interstate Route 84 as it ascends from the floor of the Delaware River valley to the uplands). This fault plane parallels bedding and strikes N62°E and dips 13°N. Slickensides on the fault surface are oriented N45°W. The fault is offset at one place where the plane of movement changed bedding surfaces. The offset zone is about 6 feet long and 2 feet thick, brecciated, and cemented with quartz and barite. The fault is not shown on Plate 1 because of its bedding-parallel nature.

The longest known fault in Pike County occurs north of Milford in the northeastern quarter of the Milford quadrangle (Plate 1). This fault, which trends north-northeast, is interpreted from geomorphic and stratigraphic evidence. The long ravine through which Deep Brook flows is one of the straightest and narrowest stream courses in the county. Although the ravine may have been eroded and deepened by meltwater during deglaciation, its length and depth indicate a major fracture zone in the bedrock. The ravine does

Figure 89. Fourth-order fold in sandstone of the Delaware River Member of the Catskill Formation exposed on the east side of the Erie-Lackawanna Railroad track 0.4 mile south of the railroad overpass in Shohola at 41°28′06″N/74°54′47″W.

not compare with other landscape forms in northeastern Pennsylvania that have resulted solely or partly from glacial erosion. Angular fragments of vein quartz, which are frequently associated with major fractures or faults, occur within the ravine. Close comparison of outcrops along Interstate Route 84 reveals that the exposure in the large cut along the eastbound lane just south of Deep Brook is in the lower part of the Trimmers Rock Formation, whereas the exposures in the cuts just northwest of Deep Brook are in the upper part of the Mahantango Formation. This sequence is a reversal of the normal succession and substantiates the presence of a fault. Vertical displacement along the fault is estimated to be 800 feet in the vicinity of Interstate Route 84; there is no displacement 2.5 miles to the north. The fault is interpreted to be a steeply north-dipping reverse fault. As such, the fault is antithetic (dipping in a direction opposite from the direction of compressional stress in the region).

Other unmapped faults occur in several places. Small near-vertical faults having displacements of less than 3 feet occur in exposures of the Mahantango Formation in some of the shale-chip-rubble borrow pits along U.S. Route 209. These small faults are not well exposed and cannot be traced through a single outcrop. The strike of these faults parallels that of adjacent joints, and the displacements may be a result of slump blocks that cannot be readily defined.

## JOINTS

Data were collected for 1,266 joints in Pike County. The strike and dip orientations of these joints are shown on Plate 1, and the data are summarized in Figure 90. There are two dominant joint sets in the county: one set has an average orientation of N6°E, and the other set has an average orientation of N85°E.

Ninety-five percent of the joints in the N6°E joint set are oriented between N4°W and N21°E. Ninety-five percent of the joints in this set have dips greater than 72 degrees, and 80 percent have dips greater than 85 degrees. Both east and west

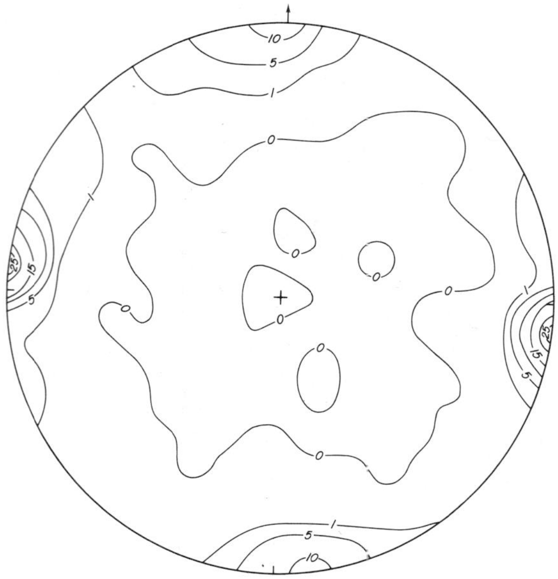

Figure 90. Contoured equal-area lower hemisphere projection of 1,266 poles to joint surfaces in Pike County. The contours are at 0, 1, 5, 10, 15, 20, and 25 percent per 1-percent area.

dips occur, but the majority are to the east. The joints in this set that occur in sandstone commonly have a spacing of 1 to 5 feet. Spacings of 6 to 10 feet and greater than 10 feet are less common and occur in about equal proportions. Spacing is very irregular in some outcrops and ranges from a few inches to a few feet. Most of the joint surfaces are smooth and planar through the extent of the outcrop (Figures 91 and 92). In many places, the joints are open at the surface. They become tight but persistent with depth (as seen in flagstone quarries and roadcuts). In siltstone and claystone, the joints occur about equally as irregularly shaped surfaces and smooth planes and generally have spacings of less than 5 feet.

Ninety-five percent of the joints in the N85 °E joint set are oriented between N70 °E and S82 °E. Ninety-five percent of the joints in this set have dips greater than 69 degrees, and 90 percent have dips greater than 85 degrees. Both north and south dips occur, but the majority are to the south. The joints in this set that occur in sandstone are equally likely to have close spacings of less than 1 foot or variable spacings between 1 and 5 feet. Frequently the spacing is highly variable within a single outcrop. Most of the joint surfaces are irregular in shape (Figure 93) and commonly represent only a parting tendency, which may or may not be persistent throughout a single outcrop. The joints from this set tend to be less open at the surface than the joints from the other set. They also become tighter with depth and appear to be less persistent with depth than the other set. In siltstone and claystone, these joints generally have irregularly shaped surfaces and spacings between 1 and 3 inches. Irregular spacing and regular spacing between 1 and 4 feet occurs in some outcrops.

Most of the joints observed in Pike County are not sealed with secondary mineralization, but white quartz up to a few millimeters thick does occur locally. The pattern of jointing appears to be consistent throughout the county; however, a larger spread of orientations occurs in the siltstone and claystone of the Mahantango Formation along the southeastern margin of the county. Joints developed in the Mahantango also show considerable variation in lateral and vertical persistence (Figure 73). All of the joints are planes of weakness along which rock disintegration can occur. Many joint surfaces have been exposed for tens of feet by glacial erosion, and numerous exposures show block separation along joint planes (Figure 44). The spacing of joints is critical in flagstone quarrying because it controls the maximum size of the flagstones.

## CLEAVAGE

With the exception of a few scattered occurrences in the red claystone of the Catskill Formation, cleavage development in Pike County is restricted to the claystone of the Mahantango Formation. Strike and dip orientations for 83 cleavage measurements indicate that the average orientation is N74 °E, 64 °S (Figure 94). Eighty-five percent of the measured cleavages are oriented between N62 °E and N86 °E and have dips between 51 °S and 78 °S. The cleavage spacing is

Figure 91. Smooth-surfaced joint exposed on an outcrop of crossbedded sandstone of the Delaware River Member of the Catskill Formation. The site is located on the south side of Snow Hill Road about 0.4 mile south of Twelvemile Pond at 41°10′06″N/ 75°05′37″W. The joint surface is oriented N12°E and dips 85°SE. The scale is 1 foot long.

generally 1 to 2 inches and is properly called spaced cleavage (Figure 95). Closer spacing occurs locally in some of the cliff exposures of the Mahantango Formation along the northwest side of the Delaware River valley. Spaced cleavage is uncommon in the claystone of the Catskill Formation in Pike County, but the cleavage is similar in orientation to that in the Mahantango Formation.

## FRACTURE TRACES

Fracture traces are the intersections of fractures or fracture zones with the surface of the earth. Many major fracture traces appear as linear features on aerial photographs. These linear features, which represent zones of extensive fracturing, may mark areas of increased groundwater supplies. Five hundred sixty-seven fracture traces were mapped in Pike County utilizing color aerial photographs. None of the interpreted fracture traces were checked in the field. Traces less than 0.25 mile in length are termed "minor fracture traces." Fracture traces may be useful in locating water wells, avoiding unsuitable sites for sanitary landfills, and estimating general construction conditions in the area.

The orientations of the fracture traces are shown in a rose diagram, which indicates a large set of traces between N15°W and N25°E and a smaller but well-developed set between N75°E and S75°E (Figure 96). These orientations correspond almost exactly to those of measured joints in the county (Figure 90).

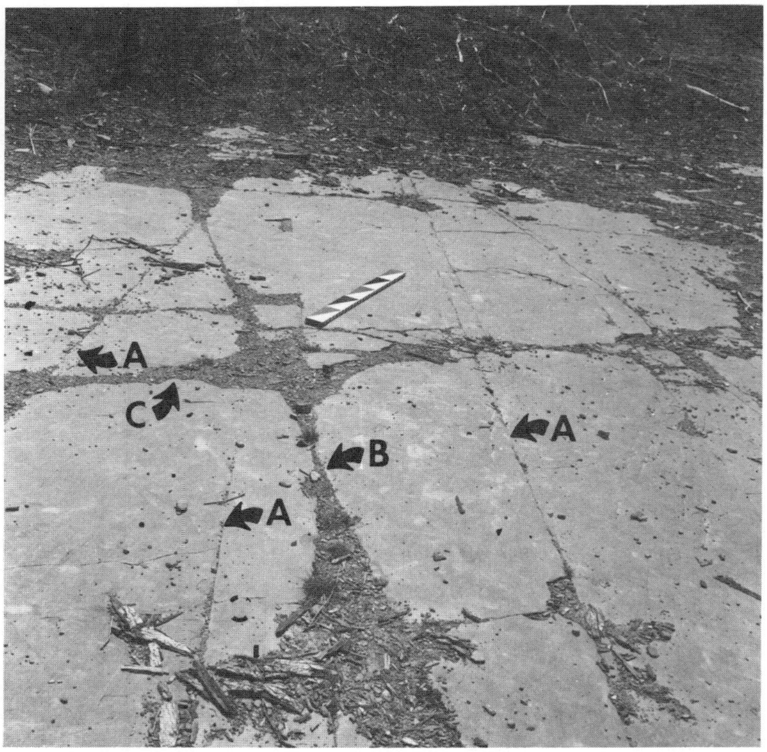


Figure 92. Joints exposed on a glaciated surface of sandstone of the Delaware River Member of the Catskill Formation. The site is about 0.55 mile east of Pa. Route 402 and 2.2 miles south of Twelvemile Pond at 41°08′46″N/75°04′17″W. The glacial striae are oriented N34°E and are parallel to the scale, which is 5 feet long. The joint orientations are as follows: A—strike N22°E, dip 80°SE; B—strike N8°E, dip 80°SE; and C—strike N80°E, dip 90°.

# MINERAL RESOURCES

## PEAT

There is a large potential for harvesting of peat for commercial purposes in Pike County. In 1981, 25,000 tons of peat (total value $647,000) was produced in Pennsylvania, mainly for humus and agricultural purposes (Kebblish and Tuchman, 1983). Although commercial peat operations existed in some nearby counties, no peat production was reported for Pike County.

The primary use of peat in the United States is as a soil conditioner. Peat has very little fertilizing capability, but tends to bind together sandy soils and make clayey soils more porous. Because of the great variation in the physical properties of peat, commercial peat must meet the regulations established by the Federal Trade Commission in 1950 governing its sale. These regulations define peat as any partially decomposed plant matter that has accumulated underwater or in a water-saturated environment. In addition to the Federal Trade Commission regulations, purchases of peat by the federal government are controlled by Federal Specification Q–P–166e (Cameron, 1970). The regulations for some of the physical properties of peat are as follows:

(1) The percentage of organic matter should not be less than 75 percent.

(2) The minimum water-holding capacity (Federal Specification only) should be 800 percent for sphagnum-moss peat, 400 percent for other moss peat, 200 percent for

Figure 93. Irregularly shaped joint surface exposed on an outcrop of conglomeratic sandstone of the Lackawaxen Member of the Catskill Formation. The site is located at the north end of White Deer Lake, about 5 miles southeast of Hawley at 41°24′39″N/75°07′55″W. The joint surface is oriented approximately east-west.

humus peat, and 400 percent for reed-sedge peat.

(3) The pH values (Federal Specification only) should be 3.2 to 4.5 for sphagnum-moss peat, 3.2 to 7.0 for other moss peat, and 4.0 to 7.5 for humus and reed-sedge peat.

The quantity of peat present in Pike County can be estimated only within generalized limits. The work of Edgerton (1969) and Cameron (1970) indicates that, in northeastern Pennsylvania, the thickness of peat deposits covering 10 to 100 acres ranges from 5 to 16 feet, but thicker deposits may occur. Areal measurement of the peat shown on Plate 2 indicates a total surface area of 18.5 $mi^2$. Using the thicknesses given above, the amount of peat present in Pike County ranges from 95,679,701 to 306,175,040 $yd^3$ (cubic yards). If an average thickness of 7.7 feet is assumed, then there are 147,346,740 $yd^3$ of peat. Based on data from Cameron (1970), an average organic content of 78.5 percent can be assumed for the peat in Pike County, although the actual organic content may range from 35 to 99 percent. Cameron also indicated that, on the average, about 91 percent of a given volume of peat in any deposit is water. Estimating the density of wet peat at 1,890 $lb/yd^3$ (pounds per

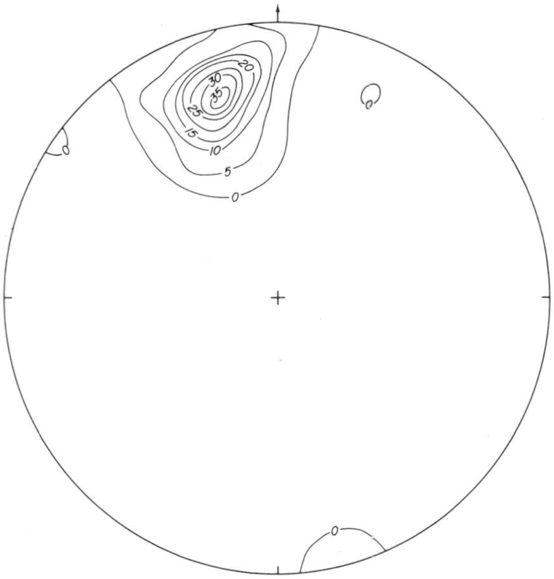


Figure 94. Contoured equal-area lower hemisphere projection of 83 poles to cleavage surfaces in Pike County. The contours are at 0, 5, 10, 15, 20, 25, 30, and 35 percent per 1-percent area.

Figure 95. Spaced cleavage in claystone of the Mahantango Formation. The cleavage is oriented N75°E, dips 60°S, and has a spacing of 1 to 2 inches. The site is located on the west side of a road about 1.3 miles north of Bushkill at 41°06′45″N/75°00′08″W. The bedding is oriented N66°E and dips 17°N, approximately parallel to the scale. The scale is 1 foot long.

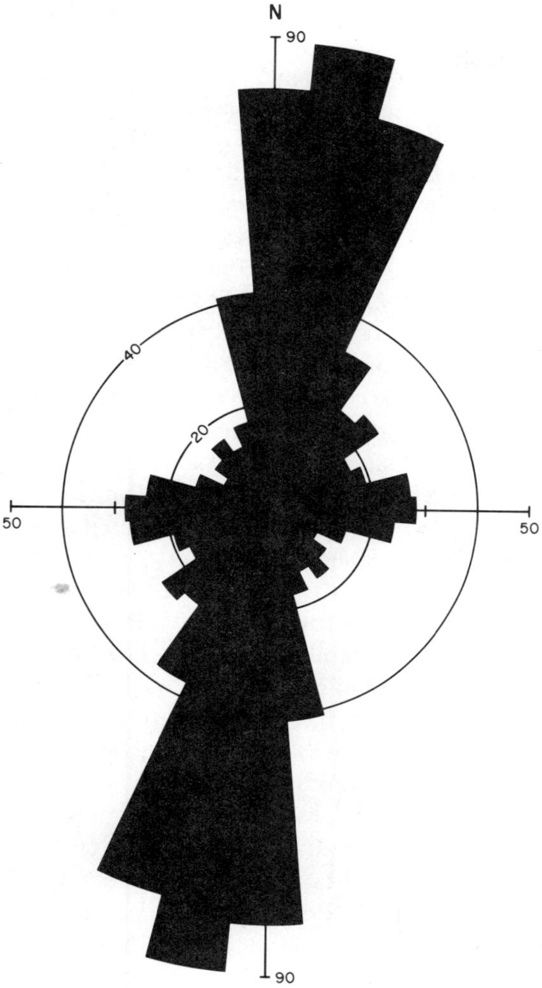


Figure 96. Rose diagram of 567 fracture traces in Pike County. The fracture traces are grouped in 10-degree intervals. The circles indicate the number of traces.

cubic yard), a cubic yard of peat would contain only 134 pounds of organic material when dry, and the total reserve tonnage of peat in Pike County, assuming an average thickness of 7.7 feet, would be nearly 10 million short tons.

In Pike County, numerous peat deposits contain the minimum 100,000 short tons of harvestable, air-dried peat needed for a commercial operation. Cameron (1970, p. A44–A45) made several suggestions for finding exploitable peat deposits. These suggestions are summarized as follows:

(1) The peat having the highest organic content and lowest ash content occurs in bogs, not along the main drainage of through-flowing streams subject to flooding.
(2) Better deposits occur where the surrounding drainage basin is not more than 10 times the size of the bog.
(3) A minimum of decomposition occurs in deposits where groundwater fluctuation has been slight. Such deposits are characterized by an absence of outlet streams, the presence of ponds, an outlet stream flowing over bedrock, or a lack of through-flowing streams.
(4) Peat deposits less than 3 feet thick or less than an acre or two in area generally have a high ash content.
(5) The thickest peat deposits having the highest organic contents tend to occur adjacent to steep bedrock slopes.

Careful study of Plate 2 will guide interested persons to prospective commercial peat deposits. Once a prospective deposit is selected, a systematic probing of the deposit should be done along several traverses to obtain an accurate evaluation of its thickness, and samples of the peat should be taken at several places and depths for testing and evaluation of its quality.

Commercial production of peat involves four operational phases: clearing, draining, harvesting, and processing. Clearing a peat bog involves the removal of all trees, stumps, and shrubs in order to prepare an operable surface. Because peat has a low bearing strength, this is best accomplished in winter when the ground is frozen.

Draining is necessary to change the water-saturated condition of the peat. This makes the bog more compact and allows equipment to move on the bog surface. Draining also reduces the weight of the handled material. Draining is best accomplished by ditching, generally using a dragline. Pumping is possible, but the cost is generally prohibitive. More detailed information on drainage is presented by Edgerton (1969, p. 39–41). Draining a bog introduces additional water into Pennsylvania waterways, and permission to drain must be obtained from the Department of Environmental Resources, Division of Water Quality, Harrisburg.

Harvesting involves harrowing the peat so it dries to a moisture content of about 60 percent, windrowing to facilitate collection, and transportation to stockpiles. All of this is accomplished using relatively light equipment. More detailed information is available in Edgerton (1969, p. 41–42).

Harvested peat is processed by screening, shredding, weighing, and packaging. The prepared product is then ready for marketing, which is the most critical factor for Pike County peat. Peat produced in Pike County must compete on the market with peat from adjacent Pennsylvania counties, northern New Jersey, nearby New England states, some midwestern states (principally Michigan), and Canada. Consequently, a careful analysis of the market potential should be undertaken prior to initiation of any commercial operation. More information on processing and marketing is presented by Edgerton (1969, p. 42–46).

## SAND AND GRAVEL

Olean ice-contact stratified sand and gravel is a potential source of large quantities of sand and gravel in Pike County. The surface area of the deposits shown on Plate 2 is calculated at 22.7 $mi^2$. Although the thickness of these deposits is variable and generally unknown, an average thickness of 50 feet seems reasonable. Using these figures, there was approximately 1,171,925,300 $yd^3$ of sand and gravel in Pike County prior to removal of some of this material. Assuming an average weight of 3,000 $lb/yd^3$ for sand and gravel, there is about 1,757,888,000 short tons of sand and gravel in the ice-contact deposits.

As previously discussed in the section "Olean Ice-Contact Stratified Sand and Gravel" (p. 66–75), the texture of these deposits is extremely variable. Size analyses of sand and gravel deposits from various localities (Figure 62) in the county are shown in Table 6. Although these analyses may be indicative of the general distribution of sizes present in the smaller size range, most deposits have a moderate to high proportion of coarser sizes, including some boulders not suitable for crushing. Size-analysis data for sand-sized material are presented in Table 7. The data in Tables 6 and 7 are not adequate to properly evaluate any deposit for commercial purposes. The presence of coarse clasts and the use of crushing and screening in commercial sand and gravel operations probably will allow most of the ice-contact stratified sand and gravel deposits to be utilized as sources of coarse and fine aggregate.

The lithologic composition of clasts from these deposits was previously discussed in the section "Olean Ice-Contact Stratified Sand and Gravel," and the relationship between the underlying bedrock and the sand and gravel was demonstrated. Table 8 shows the results of standard freeze-thaw tests, which determine the extent of clast breakdown. The tested samples are arranged in relationship to underlying bedrock. Although samples 25, 26, and 27 are not underlain by rocks of the Poplar Gap and Packerton Members, the deposits from which the samples were taken have a considerable amount of rock derived from those members because of their close proximity to the Poplar Gap and Packerton outcrop. The table shows that the clasts from samples related to the underlying Poplar Gap and Packerton Members had appreciably greater breakdown during freeze-thaw testing than did other samples. Examination of the tested samples revealed that this resulted from disintegration of the coarse-grained rocks. In all of the samples, breakup was most intense in the more friable, coarser grained rocks, particularly intraformational conglomerates. Durability of claystones was generally good except where they appeared to be weathered. Finer grained sandstone and siltstone seemed to be the most resistant to breakup. Most of the breakup was by granular disintegration rather than platy. The amount of breakdown in the samples has little correlation (correlation coefficient is 0.47) with absorption (Table 8).

In summary, a large quantity of sand and gravel exists in Pike County in the Olean ice-contact deposits. If processed by crushing and

Table 6. *Size-Analysis Data for Sand and Gravel Samples Collected from Deposits of Olean Ice-Contact Stratified Sand and Gravel in Pike County*[1]

(Values represent percentage of sample passing sieve size)

| Sample number | Sieve size (inches) | | | | | Sieve size (number[2]) | | | | | | |
|---|---|---|---|---|---|---|---|---|---|---|---|---|
| | 2 | 1½ | ¾ | ½ | ⅜ | 4 | 8 | 16 | 30 | 50 | 100 | 200 |
| 1 | 93.50 | 84.08 | 72.41 | 65.38 | 59.81 | 44.69 | 25.03 | 11.17 | 4.92 | 2.68 | 2.23 | 1.79 |
| 2 | 79.13 | 69.78 | 50.00 | 40.19 | 33.80 | 17.29 | 6.22 | 1.73 | .86 | .69 | .52 | .35 |
| 3 | 92.27 | 87.90 | 64.71 | 42.35 | 26.05 | 8.24 | 4.28 | 2.31 | 1.48 | 1.07 | .82 | .66 |
| 4a | 75.39 | 62.37 | 43.67 | 36.34 | 31.84 | 23.97 | 17.37 | 11.51 | 7.19 | 4.31 | 2.64 | 1.92 |
| 4b | 86.78 | 77.07 | 62.19 | 55.99 | 51.86 | 41.94 | 32.71 | 26.00 | 22.65 | 18.87 | 12.16 | 5.45 |
| 5 | 77.16 | 70.53 | 57.46 | 51.75 | 47.88 | 38.31 | 27.58 | 13.03 | 3.83 | 1.53 | .77 | .38 |
| 6 | 94.52 | 89.46 | 70.18 | 57.46 | 48.68 | 32.02 | 23.37 | 16.97 | 11.85 | 6.72 | 3.20 | 1.28 |
| 7 | 100.00 | 99.25 | 86.57 | 77.61 | 70.15 | 55.22 | 46.38 | 38.10 | 29.82 | 21.54 | 10.49 | 3.86 |
| 8 | 100.00 | 99.15 | 90.35 | 81.89 | 74.79 | 61.25 | 53.90 | 48.39 | 42.26 | 28.17 | 9.80 | 3.06 |
| 9 | 93.68 | 85.15 | 70.14 | 63.35 | 58.45 | 48.02 | 39.86 | 32.17 | 25.45 | 16.81 | 10.56 | 6.72 |
| 10 | 91.24 | 81.71 | 59.53 | 49.61 | 43.77 | 33.46 | 25.09 | 17.40 | 9.37 | 4.01 | 1.67 | 1.00 |
| 11 | 90.00 | 85.26 | 67.54 | 58.77 | 53.16 | 40.35 | 29.45 | 19.77 | 12.91 | 6.86 | 3.23 | 2.42 |
| 12a | 97.66 | 90.65 | 64.25 | 49.77 | 41.82 | 29.44 | 21.79 | 15.60 | 11.48 | 7.36 | 3.24 | 1.18 |
| 12b | 100.00 | 100.00 | 96.99 | 94.65 | 92.64 | 85.95 | 74.78 | 53.29 | 29.22 | 12.89 | 3.44 | .86 |
| 13 | 90.71 | 79.90 | 64.02 | 55.41 | 50.51 | 40.88 | 34.75 | 30.25 | 25.35 | 13.89 | 4.50 | 1.64 |
| 14 | 94.09 | 81.80 | 58.63 | 48.70 | 42.32 | 31.21 | 25.60 | 21.85 | 17.48 | 8.43 | 3.12 | 1.56 |
| 15 | 88.76 | 80.87 | 68.44 | 63.12 | 59.96 | 54.44 | 51.72 | 49.54 | 44.10 | 21.23 | 6.53 | 2.18 |
| 16 | 90.44 | 82.27 | 55.98 | 44.02 | 37.45 | 27.09 | 20.86 | 16.25 | 11.64 | 5.96 | 4.44 | 1.35 |
| 17 | 89.61 | 75.27 | 53.76 | 46.06 | 40.86 | 31.00 | 25.11 | 20.46 | 15.50 | 10.23 | 7.44 | 5.27 |
| 18 | 83.74 | 73.53 | 55.20 | 47.07 | 42.16 | 32.33 | 22.63 | 14.55 | 9.05 | 6.14 | 3.56 | 2.26 |
| 19 | 96.03 | 93.21 | 76.99 | 67.22 | 59.77 | 45.53 | 37.79 | 27.32 | 13.66 | 4.55 | 1.82 | 1.37 |
| 20 | 90.98 | 78.08 | 53.96 | 45.44 | 40.88 | 30.75 | 21.52 | 13.53 | 7.99 | 3.38 | 1.23 | .61 |
| 21 | 90.84 | 86.01 | 72.67 | 64.79 | 58.84 | 43.89 | 34.23 | 26.33 | 19.31 | 10.53 | 3.07 | .88 |
| 22 | 75.61 | 67.90 | 47.79 | 41.37 | 37.80 | 30.67 | 23.62 | 13.80 | 6.13 | 2.45 | 1.23 | .61 |
| 23a | 89.21 | 80.32 | 58.10 | 49.21 | 42.54 | 28.25 | 16.38 | 6.78 | 2.82 | 1.69 | 1.13 | .85 |
| 23b | 100.00 | 95.85 | 83.68 | 79.23 | 75.96 | 65.58 | 51.81 | 34.10 | 13.12 | 2.62 | 1.31 | .66 |
| 24 | 100.00 | 97.07 | 87.53 | 79.46 | 73.11 | 60.64 | 52.15 | 45.48 | 40.02 | 30.32 | 17.59 | 10.31 |
| 25 | 93.39 | 86.78 | 77.14 | 72.73 | 70.80 | 65.84 | 62.55 | 60.57 | 58.60 | 55.31 | 45.43 | 25.68 |
| 26 | 91.11 | 83.33 | 59.17 | 46.67 | 38.89 | 23.89 | 16.96 | 15.05 | 13.86 | 9.79 | 5.02 | 3.11 |
| 27 | 85.82 | 77.86 | 53.48 | 43.78 | 39.05 | 31.84 | 27.06 | 24.20 | 22.29 | 17.83 | 9.55 | 3.82 |
| 28a | 90.62 | 86.98 | 70.14 | 59.90 | 52.95 | 40.28 | 33.03 | 27.79 | 21.35 | 9.67 | 3.22 | .81 |
| 28b | 100.00 | 98.80 | 96.86 | 94.93 | 93.48 | 86.71 | 80.64 | 71.10 | 56.36 | 29.48 | 8.67 | 2.60 |

[1]Sample locations are shown in Figure 62.

[2]The size of sieve numbers 4, 8, 16, 30, 50, 100, and 200 are 4.76, 2.38, 1.19, 0.595, 0.297, 0.149, and 0.074 mm, respectively.

screening, most of these deposits contain an adequate distribution of grain sizes to be used for production of various coarse and fine aggregate. Some deleterious components exist, but crushing would probably reduce the already small quantity of these clasts. In general, the quality of the various lithologies is good. One problem with the ice-contact deposits is that many individual deposits are small and not suitable for long-term, large quarrying operations.

Outwash sand and gravel deposits in Pike County have an even larger reserve of sand and gravel of presumably high quality. These deposits have not been evaluated because land use and potential environmental concerns related to quarrying operations in valley bottoms and near large

Table 7. *Size-Analysis and Fineness-Modulus Data for Sand Samples Collected from Deposits of Olean Ice-Contact Stratified Sand and Gravel in Pike County*[1]

| Sample number | Sieve size[2, 3] (number) | | | | | | | Loss by wash[4] (percent) | Fineness modulus |
|---|---|---|---|---|---|---|---|---|---|
| | 4 | 8 | 16 | 30 | 50 | 100 | 200 | | |
| 1 | 100 | 56 | 25 | 11 | 6 | 5 | 4 | 4.4 | 3.97 |
| 2 | 99 | 35 | 10 | 5 | 4 | 3 | 2 | 2.0 | 4.44 |
| 3 | 100 | 52 | 28 | 18 | 13 | 10 | 8 | 6.9 | 3.79 |
| 4a | 100 | 74 | 48 | 30 | 18 | 11 | 8 | 8.3 | 3.19 |
| 4b | 100 | 78 | 62 | 54 | 45 | 29 | 13 | 9.6 | 2.32 |
| 5 | 100 | 72 | 34 | 10 | 4 | 2 | 1 | .8 | 3.78 |
| 5b | 100 | 100 | 100 | 100 | 100 | 95 | 58 | 41.9 | .05 |
| 6a | 100 | 100 | 100 | 97 | 58 | 10 | 2 | 3.0 | 3.06 |
| 6b | 100 | 73 | 53 | 37 | 21 | 10 | 4 | 1.3 | 1.35 |
| 7 | 100 | 84 | 69 | 54 | 39 | 19 | 7 | 5.7 | 2.35 |
| 8 | 100 | 88 | 79 | 69 | 46 | 16 | 5 | 3.9 | 2.02 |
| 9 | 100 | 83 | 67 | 53 | 35 | 22 | 14 | 11.8 | 2.40 |
| 10 | 100 | 75 | 52 | 28 | 12 | 5 | 3 | 2.6 | 3.28 |
| 11 | 100 | 73 | 49 | 32 | 17 | 8 | 6 | 5.1 | 3.21 |
| 12a | 100 | 87 | 62 | 34 | 15 | 4 | 1 | 3.4 | 2.98 |
| 12b | 100 | 74 | 53 | 39 | 25 | 11 | 4 | .5 | 2.98 |
| 13 | 100 | 85 | 74 | 62 | 34 | 11 | 4 | 3.9 | 2.34 |
| 14a | 100 | 82 | 70 | 56 | 27 | 10 | 5 | 4.2 | 2.55 |
| 14b | 100 | 100 | 99 | 96 | 69 | 27 | 11 | 8.3 | 1.09 |
| 15 | 100 | 95 | 91 | 81 | 39 | 12 | 4 | 3.3 | 1.82 |
| 16a | 100 | 100 | 100 | 100 | 91 | 38 | 10 | 4.8 | 2.89 |
| 16b | 100 | 77 | 60 | 43 | 22 | 9 | 5 | 6.4 | .71 |
| 17 | 100 | 81 | 66 | 50 | 33 | 24 | 17 | 13.8 | 2.46 |
| 18 | 100 | 70 | 45 | 28 | 19 | 11 | 7 | 6.5 | 3.27 |
| 19 | 100 | 83 | 60 | 30 | 10 | 4 | 3 | 2.5 | 3.13 |
| 20a | 100 | 70 | 44 | 26 | 11 | 4 | 2 | 2.0 | 3.45 |
| 20b | 100 | 99 | 96 | 85 | 45 | 10 | 3 | 2.0 | 1.65 |
| 21 | 100 | 78 | 60 | 44 | 24 | 7 | 2 | 1.8 | 2.87 |
| 22 | 100 | 77 | 45 | 20 | 8 | 4 | 2 | 2.0 | 3.46 |
| 23a | 100 | 79 | 52 | 20 | 4 | 2 | 1 | 3.2 | 3.98 |
| 23b | 100 | 58 | 24 | 10 | 6 | 4 | 3 | 1.5 | 3.43 |
| 24 | 100 | 86 | 75 | 66 | 50 | 29 | 17 | 15.0 | 1.94 |
| 25 | 100 | 95 | 92 | 89 | 84 | 69 | 39 | 28.4 | .71 |
| 26a | 100 | 71 | 63 | 58 | 41 | 21 | 13 | 12.5 | 2.46 |
| 26b | 100 | 100 | 99 | 94 | 55 | 11 | 2 | 1.3 | 1.41 |
| 27 | 100 | 100 | 100 | 100 | 97 | 74 | 33 | 7.8 | 1.83 |
| 27b | 100 | 85 | 76 | 70 | 56 | 30 | 12 | 22.6 | .29 |
| 28a | 100 | 82 | 69 | 53 | 24 | 8 | 2 | 2.1 | 2.64 |
| 28b | 100 | 100 | 100 | 99 | 91 | 52 | 16 | 2.7 | 2.16 |
| 28c | 100 | 93 | 82 | 65 | 34 | 10 | 3 | 1.4 | .58 |

[1]Sample locations are shown in Figure 62.
[2]See Table 6 for millimeter size of each sieve number.
[3]Values represent percentage of sample passing sieve size.
[4]Silt and clay present in original sample and removed by washing prior to sieving.

Table 8. *Breakdown, Absorption, Specific-Gravity, and Composition Data for Pebbles Retained on a ¾-Inch Sieve for Selected Samples Collected from Deposits of Olean Ice-Contact Stratified Sand and Gravel in Pike County*[1]

| | | | | Composition[2] | | | | | | | | |
|---|---|---|---|---|---|---|---|---|---|---|---|---|
| Sample number | Breakdown[3] (percent) | Absorption (percent) | Specific gravity | Conglomerate and conglomeratic sandstone | Very coarse to coarse-grained sandstone | Medium-grained sandstone | Fine-grained red sandstone | Nonred siltstone | Red siltstone | Nonred claystone and shale | Red claystone and shale | Intraformational conglomerate |
| TRIMMERS ROCK AND MAHANTANGO FORMATIONS[4] | | | | | | | | | | | | |
| 1 | 6.77 | 0.99 | 2.67 | 1 | 1 | — | — | 45 | — | 1 | — | — |
| 2 | 1.26 | 1.70 | 2.60 | 1 | — | — | — | 22 | — | 5 | — | 1 |
| 3 | 6.47 | 1.05 | 2.65 | — | — | — | — | 19 | — | 9 | — | — |
| 4a | 6.14 | 1.03 | 2.64 | — | 1 | 1 | — | 15 | — | — | — | — |
| 4b | .34 | 1.01 | 2.64 | — | — | — | — | 10 | — | 7 | — | 2 |
| TOWAMENSING MEMBER OF CATSKILL FORMATION[4] | | | | | | | | | | | | |
| 5 | 2.17 | 1.09 | 2.63 | — | — | 7 | — | 1 | 2 | 1 | — | — |
| 6 | 4.63 | 1.24 | 2.63 | — | — | — | — | 4 | — | 5 | — | 1 |
| LACKAWAXEN AND DELAWARE RIVER MEMBERS OF CATSKILL FORMATION[4] | | | | | | | | | | | | |
| 7 | 6.34 | 1.32 | 2.62 | — | — | 2 | — | 3 | 1 | 1 | — | — |
| 8 | 7.64 | 1.29 | 2.63 | 3 | — | 5 | 1 | 1 | — | — | 1 | 2 |
| 9 | 2.54 | 1.26 | 2.65 | — | — | — | — | 2 | 4 | 1 | — | 2 |
| 10 | .25 | 1.16 | 2.65 | 1 | — | — | 4 | 4 | 6 | — | 4 | — |
| 11 | 7.62 | 1.25 | 2.64 | 1 | — | 3 | — | 1 | 3 | 1 | — | — |
| 12 | 6.91 | 1.30 | 2.65 | — | — | 3 | 5 | 3 | 3 | 3 | 1 | — |
| 13 | 5.91 | .91 | 2.66 | 4 | — | 2 | 1 | 6 | 6 | 3 | 3 | 2 |
| 14a | 4.56 | .94 | 2.65 | 2 | — | 2 | 2 | — | 4 | 1 | — | 1 |
| 15 | 6.09 | 1.33 | 2.64 | — | — | 4 | 2 | 3 | 6 | — | 2 | — |
| 16 | .49 | .92 | 2.65 | 1 | 2 | 2 | 1 | 1 | 5 | — | — | — |
| 17 | 10.85 | 1.45 | 2.64 | 1 | — | — | 1 | — | 5 | 1 | 3 | — |
| 18 | 9.62 | 1.00 | 2.65 | 1 | — | — | — | — | 2 | — | 1 | 1 |
| 19 | 1.57 | 1.06 | 2.65 | — | 2 | 1 | 3 | — | — | — | 1 | — |
| 20 | 5.78 | 1.29 | 2.63 | — | 3 | 5 | 1 | 1 | 4 | 1 | 3 | — |
| 21 | 8.75 | .93 | 2.67 | — | 1 | — | — | 1 | 1 | 1 | — | — |
| 22 | 7.59 | .90 | 2.67 | — | — | 1 | — | 1 | 5 | 3 | 3 | 2 |
| POPLAR GAP AND PACKERTON MEMBERS, UNDIVIDED OF CATSKILL FORMATION[5] | | | | | | | | | | | | |
| 23a | 24.17 | 2.17 | 2.57 | — | 4 | 5 | 5 | 7 | — | — | — | — |
| 23b | 10.58 | 1.89 | 2.60 | 1 | 2 | — | — | 3 | 3 | 1 | — | — |
| 24 | 7.56 | 1.41 | 2.63 | 3 | — | 9 | — | — | 10 | — | — | 1 |

Table 8. *(Continued)*

| Sample number | Breakdown[3] (percent) | Absorption (percent) | Specific gravity | Composition[2] | | | | | | | | |
|---|---|---|---|---|---|---|---|---|---|---|---|---|
| | | | | Conglomerate and conglomeratic sandstone | Very coarse to coarse-grained sandstone | Medium-grained sandstone | Fine-grained red sandstone | Nonred siltstone | Red siltstone | Nonred claystone and shale | Red claystone and shale | Intraformational conglomerate |
| POPLAR GAP AND PACKERTON MEMBERS, UNDIVIDED OF CATSKILL FORMATION[5] *(Continued)* | | | | | | | | | | | | |
| 25 | 15.36 | 1.40 | 2.60 | 2 | 2 | 2 | 3 | 2 | — | — | 1 | — |
| 26 | 20.08 | 1.19 | 2.63 | 2 | — | 1 | — | 1 | 4 | — | 2 | 5 |
| 27 | 16.15 | 1.36 | 2.62 | 4 | 1 | 4 | 1 | 1 | 1 | — | 1 | — |
| 28 | 20.59 | 1.29 | 2.62 | 3 | 1 | 6 | 9 | 6 | — | — | 1 | 7 |

[1]Sample locations are shown in Figure 62.
[2]Values represent number of pebbles.
[3]Standard freeze-thaw test.
[4]Underlying bedrock.
[5]Underlying bedrock or bedrock in close proximity to samples. See text discussion.

streams make it unlikely that they will be utilized in the near future. Some of the deposits probably have greater value as a source of groundwater.

## BLUESTONE (FLAGSTONE)

An abundance of excellent-quality bluestone, a name given to bluish-gray flagstone, exists in Pike County. Bluestone is a dense, hard, fine-grained, feldspathic sandstone, which splits easily along bedding planes into smooth slabs of various thicknesses. These slabs are used for flagging, treads, veneer, coping, sills, wall stones (Figure 97), slab stock, hearthstones, and mantles. During the latter part of the nineteenth century, a large flagstone industry flourished in the county. According to Fluhr (1970), bluestone was quarried and shipped from northeastern Pennsylvania as early as 1830, but received little favor in a world that used brick. By the end of the Civil War the use of bluestone was catching on, and regular shipments were made on the Delaware and Hudson Canal and the Erie Railway.

By 1886, the Kilgour Blue-Stone Company, headquartered at Parkers Glen in Pike County, operated up to 150 bluestone quarries and employed 500 men during its peak production time in the spring (Matthews, 1886; Fluhr, 1970). The Kilgour stone mills at Parkers Glen operated 24 hours a day sawing, planing, and polishing bluestone for all types of uses, particularly flagging. The Kilgour Company was only one of several operations then active in Pike County.

During the peak years of production, the bluestone industry in Pike County produced stone considered to be of exceptional quality and beauty. Some of it was shipped as far away as Cuba for use as sidewalks. Large quantities of specific sizes were sometimes required (e.g., 100 pieces, 8 feet by 5 feet by 5 inches, for the New Jersey Prison at Newark). Some pieces of extraordinary size were also quarried and shipped (e.g., 24 feet by 15 feet by 8 inches and 22 feet by 32 feet by 8 inches, shipped to New York City). Business was good, and the Milford Dispatch reported in February 1891 that the annual bluestone product of the Delaware Valley was over $1,000,000.

Figure 97. Bluestone used for building stone in a former silk mill (now a dress factory) in Hawley, Pennsylvania. The stone was quarried along the east bank of Wallenpaupack Creek below the factory site.

As the nineteenth century ended, Portland cement began to replace natural stone as a construction material. A peak of bluestone production in 1912 was followed by a rapid decline and only minimal production by 1932 (Stone, 1923; 1932). There are no known active quarrying operations in Pike County today. Although the market is not large, a demand for good-quality flagstone still exists, and Pike County produced some of the best.

Bluestone occurs in the Towamensing, Delaware River, and Lackawaxen Members of the Catskill Formation, and the largest quantities were produced from the latter two members. There seems to be very little, if any, difference in the bluestone produced from the different members. Bluestone is derived from the lower, planar-stratified sandstone of three similar types of vertical sequences.

One sequence consists of planar-stratified, bluish-gray sandstone grading upward into finer grained, crossbedded sandstone (Figure 98). The lowest crossbeds are generally very low angle and tabular. Successive sets have higher angle, tabular or trough crossbedding and erosional bases. Sandstone above a disconformity is generally

Figure 98. Flagstone quarry developed in the Lackawaxen Member of the Catskill Formation 1.1 miles southeast of Kimble on Shimers Mountain at 41°27′28″N/75°06′22″W. The lower, planar-stratified sandstone grades upward into crossbedded sandstone.

coarser grained than sandstone below a disconformity. In another type of sequence, planar-bedded, bluish-gray sandstone is separated by a pronounced disconformity from an overlying coarser grained, trough-crossbedded sandstone (Figure 99). In the third sequence, planar-bedded, bluish-gray sandstone grades upward into tabular- or trough-crossbedded sandstone, which in turn grades upward into planar-bedded sandstone followed by red siltstone and claystone. The first two sequences occur with about equal frequency; the third sequence is uncommon and not present in the Towamensing Member. The three sequences have an apparently random vertical stacking.

The planar-bedded sandstones generally appear structureless in fresh outcrop, but they possess distinct laminations, which have spacings of a few millimeters or less and in some places have barely detectable reactivation surfaces. Parting into flaggy pieces 1 or more centimeters thick occurs along these laminae (Figure 100). Some bedding surfaces display primary current lineation (ties) and parting-step lineation (Figure 101). Many surfaces contain abundant mica flakes, and some surfaces exhibit finely broken plant fragments. Ripple bedding is noticeably lacking in both the planar-bedded and crossbedded sandstones. In many places, sequences of planar-bedded sandstone are less than 10 feet thick, but in some places they are up to 30 feet thick. Some planar-bedded units appear to be continuous for several hundred feet in directions parallel to the orientation of their parting lineations. The con-

Figure 99. Flagstone quarry face in the Delaware River Member of the Catskill Formation on the north side of Pond Eddy Creek about 0.8 mile west-southwest of its junction with the Delaware River at 41°25′34″N/74°50′25″W. Crossbedded sandstone unconformably overlies the planar-bedded sandstone. The scale is divided in inches and centimeters.

tinuity normal to parting lineations appears to be less; the planar beds are apparently replaced laterally by crossbeds. Crossbedded sandstones are more abundant, but apparently have more limited lateral continuity than the planar-bedded sandstones.

The microscopic composition of 68 flagstone samples from quarries in Pike County is given in Table 9. Comparison of the data in this table with the data in Tables 1, 2, and 3 (data for all sandstone types in the Towamensing, Delaware River, and Lackawaxen Members of the Catskill Formation, respectively) indicates that the planar-bedded sandstones have a higher percentage of illite-chlorite-chert cement. The long axes of all particles in the flagstone samples are well aligned parallel to both lamination and current-flow direction (as indicated by parting lineations). Heavy minerals almost always occur as concentrations along individual laminae and are mainly zircons. Coarse micas, both muscovite and biotite, are scattered throughout and are also concentrated along single laminae. Field observation of many flagstone pieces suggests that natural and artificial separation of the planar-bedded sandstones occurs most readily along laminae having high mica concentrations because these laminae have weaker binding than laminae lacking concentrations of coarse mica.

Many of the over 100 inactive quarries examined contain sufficient amounts of planar-bedded sandstone for additional quarrying, and all of the quarries have large waste piles that have some potential as a source of crushed rock and random-sized flagstone (Figure 102). Two problems are apparent in some of the larger quarries. First,

Figure 100. Natural separation of flaggy beds in a flagstone quarry developed in the Lackawaxen Member of the Catskill Formation (see Figure 98 for the location). The short length of the square is about 1 foot long.

Figure 101. Parting-step lineation (right of the scale) on the surface of a typical flagstone from the Lackawaxen Member of the Catskill Formation (see Figure 98 for the location). The scale is divided in inches and centimeters.

because most of the quarries are located on hillsides and the bluestone is mined into the hillside, overburden thickness becomes a hindrance to profitable operation. Second, as quarrying proceeds into a hillside, the bluestone becomes fresher and can be split only into thicker and thicker pieces. This limits the number of uses for the bluestone and may have been a factor in closing some quarries. The development of planes of splitting (rifts) results from decreased adhesion between the laminae in a bluestone and is caused by weathering and time-related pressure release following removal of overburden. As quarrying proceeds and fresher rock is mined, the bluestone splits only along the weakest planes (reeds), which are the planes that coincide with significant changes in texture or mineralogy. Joints, which aid quarrying and also limit stone size, are well developed in Pike County and generally have a spacing of 10 feet or more, although irregularly spaced and oriented joints (cutters) are not uncommon.

Excellent-quality bluestone awaits future development in Pike County. Although quarrying bluestone involves much hand operation, improved cutting and drilling equipment can greatly assist quarry operations. Almost all of the existing quarries once had road access, which could be reestablished with moderate ease. Knowledge of the relationship between parting lineation and continuity of planar stratification should allow quarry development in the direction of greatest bluestone persistence.

Table 9. *Microscopic Composition of Pike County Flagstones (68 Samples)*

(Values represent percentage of total composition unless otherwise indicated)

| Component | Towamensing Member (7 samples) | Delaware River Member (28 samples) | Lackawaxen Member (33 samples) |
|---|---|---|---|
| Framework | 63 | 63 | 63 |
| Cement | 35 | 34 | 35 |
| Matrix | 2 | 3 | 2 |
| Framework | | | |
| Strained quartz | 54 | 48 | 48 |
| Feldspar | 1 | 3 | 5 |
| Microcrystalline quartz | Tr[1] | 1 | Tr |
| Polycrystalline quartz | Tr | 1 | 1 |
| Stretched metaquartz | Tr | Tr | Tr |
| Shale clasts | 6 | 5 | 4 |
| Siltstone clasts | 0 | Tr | Tr |
| Metamorphic rock fragments | Tr | 2 | 1 |
| Coarse micas | 2 | 3 | 4 |
| Miscellaneous | Tr | Tr | Tr |
| Cement | | | |
| Illite-chlorite-chert composite | 27 | 22 | 21 |
| Microcrystalline silica | 2 | 4 | 7 |
| Authigenic silica overgrowths | 6 | 6 | 7 |
| Calcite | Tr | 2 | Tr |
| Matrix | | | |
| Kaolinite, illite, chlorite, and limonite | 2 | 3 | 2 |
| Average grain size (mm) | .11 | .13 | .15 |

[1]Tr, trace.

The location of bluestone quarries in Pike County was originally influenced by the proximity to transportation facilities along the Delaware and Lackawaxen Rivers and the ease of locating quarriable rock on the outcrop-abundant steep slopes adjacent to these rivers. Although sandstone suitable for quarrying exists to the southwest within the same stratigraphic units, slopes are less steep and outcrops are less abundant, which makes prospecting for suitable bluestone more difficult. Plate 1 shows the location of all identified flagstone quarries.

A quarry operation in Pike County at this time would be an economic venture of some uncertainty. Consideration should be given to the market need, the existence of a competitive flagstone industry in nearby Susquehanna County (Mikutowicz and Schenck, 1970a; Krajewski and Williams, 1971), and some of the suggestions for industry improvement in Susquehanna County by Mikutowicz and Schenck (1970b).

## SHALE-CHIP RUBBLE

Shale-chip rubble has been mined in many pits along the Delaware Valley (Figure 71) for years and has been mainly used as a fill material. Its uses include surfacing for low-use roads, backfill around foundations, and random fill to build up low areas. Tests run on this material by the Pennsylvania Department of Transportation indicate that it is suitable for shale embankments that can be compacted in loose 8-inch layers.

Tests on a sample of shale-chip rubble from adjacent Monroe County yielded the following results: bulk specific gravity, 2.61; absorption, 4.22 percent; Modified Washington Degradation Factor, 13.3; and moisture-density relationship under compaction, maximum density of 122 $lb/ft^3$ (pounds per cubic foot) with optimum moisture content of 13 percent.

Shale-chip rubble is easily mined using a front-end loader and requires no treatment once cleared of surface vegetation. The main problems associated with mining are slope failure caused by undercutting and the subsequent downslope movement of trees and other vegetation.

No records are available to indicate the amount of rubble removed from the various borrow pits, but simple estimates indicate that several million cubic feet of rubble has been mined. As much, and probably more, rubble remains as has been mined, and thus it represents a valuable resource reserve. The process of mining shale-chip rubble changes the natural environment by removing the vegetation and rubble and exposing the underlying rock surface, which is then subject to mechanical disintegration and subsequent development of more shale-chip rubble. Such mining may require (depending on tonnage) a surface-mining permit, a mine-drainage permit, and an operators permit, all of which are obtainable from the Department of Environmental Resources. However, much of the shale-chip rubble in Pike County is now part of the Delaware

Figure 102. Waste pile at a flagstone quarry. The amount of waste is typical of the large amounts found at all quarries in Pike County. The scale is 1 foot long.

Water Gap National Recreation Area and is presumably not available for use.

## CRUSHED ROCK

Crushed rock is used as railway ballast, road base course, rip rap, drainage fill material, and coarse aggregate for concrete. Crushed rock has an angular particle shape, and its limiting size and size grading can be readily controlled. Although specific requirements for any crushed rock depend upon its use, the general requirements for a satisfactory aggregate can be applied to all crushed rock. To meet these requirements, the rock (1) must be clean; (2) must be correctly graded in size; (3) must be uniform in grading and quality; (4) must be durable; and (5) must not possess any mineralogic or chemical characteristics that will cause it to react with Portland cement after mixing and placing. The first three requirements are met readily by proper processing techniques. Requirements 4 and 5 depend upon the lithologic and mineralogic characteristics of the source rock, and failure to properly assess these characteristics may result in an undesirable aggregate.

At present, crushed rock is produced at only one location in Pike County, northwest of Milford. The crushed rock there is quarried by the Eureka Crushed Stone Company from gray, crossbedded sandstones of the Towamensing Member of the Catskill Formation (Figure 15). These sandstones are good in quality, possess relatively thick beds, and are cut by numerous fractures, which aid fragmentation when blasted. Similar characteristics are found in the Towamensing Member across the southeastern length of the county.

All of the sandstones in the Delaware River, Lackawaxen, and Poplar Gap and Packerton Members have potential as crushed rock. Some of these sandstones may be less suitable than the sandstones in the Towamensing because of their flaggy character and the presence of abundant micaceous minerals on many parting surfaces, but their durability should be good. Much rock exists in the waste piles of the many flagstone quarries in the northeastern part of the county near the Delaware River and along the Lackawaxen River. These rocks, if crushed to sizes that would decrease or eliminate undesirable flat faces, should make good-quality crushed rock. The Trimmers Rock and Mahantango Formations are generally less suitable for crushed rock because of the presence of large amounts of siltstone and claystone.

## CLAY SHALE AND CLAY

Claystone samples from the Catskill and Mahantango Formations were subjected to preliminary ceramic testing, and the data are presented in Appendix 4. The tests indicate that some of the grayish-red claystones of the Catskill Formation have potential as structural clay products (e.g., building brick or floor brick) if fired to 1100 to 1200 °C. However, some of the red claystones have dispersed carbonate minerals, which are deleterious to ceramic use. Therefore, each site would require careful and detailed sampling and testing prior to development for use.

All of the tested claystone samples from the Mahantango Formation have some potential for use in structural clay products, and most of the samples have marginal potential as a lightweight aggregate because they expand if heated beyond the temperature range for structural clay products. Almost all of the Mahantango samples contain some carbonate, which limits their potential.

Abundant reserves of claystone from the Mahantango Formation exist in Pike County, and careful sampling of the lower part of the formation in the area of Bushkill would probably prove the existence of adequate amounts of carbonate-free claystone. Moderately thick deposits of red claystone in the Catskill Formation occur in the western and extreme northern parts of the county. However, most of these deposits have zones containing minor to abundant amounts of carbonate, either disseminated or as nodules, and selective quarrying may not be feasible.

No samples of glacial clay from Pike County were collected and tested, but many (perhaps most) of the peat deposits in the county are underlain by light-colored plastic clays. These clays may have some potential for use as modeling clay or ceramics. They also may have considerable potential as liners for sanitary landfills, but to determine the potential for such use requires further testing and careful evaluation of the quantity of material available at any prospective site. No deposits of clay other than those beneath peat are known in Pike County.

## RED SILTSTONE AND CLAYSTONE

Red siltstone and claystone have been quarried in Pike County from the Delaware River, Lackawaxen, and Poplar Gap and Packerton Members of the Catskill Formation for use as fill material and surfacing for low-use secondary roads. The siltstone and claystone are easily quarried by tractor-ripper or front-end loader, and they break up readily into relatively small, hackly and chippy fragments and some larger fragments. The fragments pack fairly well under moderately heavy equipment to form an attractive red road surface. Normal traffic on low-use roads causes some fragment breakdown and the creation of dust and mud, but this material washes down into the interstices, and continued road use increases the packing firmness. The open framework provides good drainage until such time as the interstices become filled with surface-abraded material.

Small quarries have been opened in these red rocks at various places in Pike County and are used intermittently whenever small quantities of the rock are needed. Most of the deposits of red siltstone and claystone are only a few feet thick, but thicker deposits occur in the western and northern parts of the county.

## GAS AND OIL

At present there is no proven gas and oil potential in Pike County, and available information indicates that the Middle and Upper Devonian rocks, valuable source rocks in western Pennsylvania, probably have no potential. The potential of the older, deep rocks is unknown.

Three exploration wells, the W. Hess #1, J. Haag #1, and Forests and Waters #1 have been drilled in the county (Figure 2). In the Haag well, a smell of gas at 2,628 feet and a gas show at 3,874 feet were reported from the Trimmers Rock Formation. In the Forests and Waters well, very slight shows of oil were reported from the Towamensing Member at 3,420 feet, 3,470 feet, and 3,490 feet; a slight show of gas was reported from the Tuscarora (Shawangunk) Formation at 12,090 feet; and slight shows of gas were recorded on the gas logger from the Trenton, Black River, and Beekmantown Formations from 13,650 to 13,810 feet. The significance of the shows in these two wells is unknown. No gas or oil shows were reported from the Hess well.

Epstein and others (1977) indicated that Middle Devonian rocks in Pike County have conodont Color Alteration Index (CAI) values between 4 and 5. They indicated that oil occurs where CAI values are 2 or less and gas generally occurs where values are 2 to 3, although dry gas does occur in areas having values as high as 4. Thus, it seems reasonable to expect no oil and little or no gas from the Middle and Upper Devonian rocks of Pike County.

The surface rocks of Pike County have a shallow, northwestward dip and no folds other than simple kink folds (cross section A–A′, Plate 1). No information is currently available about the subsurface structure in Pike County. Seismic-reflection profiles to the southwest of the county indicate the presence of folds in sub-Devonian rocks that are not present in Devonian and younger rocks. In addition, sub-Devonian rocks to the east in immediately adjacent New Jersey are structurally complex. Thus, it seems reasonable that structures favorable for gas and oil accumulation may exist in the subsurface of Pike County, but there is no indication as to the probability of any gas or oil.

## REFERENCES

Alvord, D. C., and Drake, A. A., Jr. (1971), *Geologic map of the Bushkill quadrangle, Pennsylvania-New Jersey*, U.S. Geological Survey Geologic Quadrangle Map GQ–908, scale 1:24,000.

American Society of Testing and Materials (1975), *Standard method of specifying colors by the Munsell system*, 1975 Annual Book of ASTM standards, D1535–68, Part 27, Paint test for formulative products and applied coatings, p. 270–291.

Berg, T. M. (1975), *Geology and mineral resources of the Brodheadsville quadrangle, Monroe and Carbon Counties, Pennsylvania*, Pennsylvania Geological Survey, 4th ser., Atlas 205a, 60 p.

________ (1977), *Bivalve burrow structures in the Bellvale Sandstone, New Jersey and New York*, New Jersey Academy of Science Bulletin, v. 22, p. 1–5.

Berg, T. M., Edmunds, W. E., Geyer, A. R., and others, compilers (1980), *Geologic map of Pennsylvania*, Pennsylvania Geological Survey, 4th ser., Map 1, scale 1:250,000, 3 sheets.

Berg, T. M., and Sevon, W. D. (1974), *Value of color aerial photography for geologic mapping of Pike County, Pennsylvania* [abs.], Geological Society of America Abstracts with Programs, v. 6, no. 1, p. 6.

Berg, T. M., Sevon, W. D., and Bucek, M. F. (1977), *Geology and mineral resources of the Pocono Pines and Mount Pocono quadrangles, Monroe County, Pennsylvania*, Pennsylvania Geological Survey, 4th ser., Atlas 204cd, 66 p.

Bouyoucos, G. J. (1928), *The hydrometer method for studying soils*, Soil Science, v. 25, p. 365–369.

Bucek, M. F. (1971), *Surficial geology of the East Stroudsburg 7.5-minute quadrangle Monroe County, Pennsylvania*, Pennsylvania Geological Survey, 4th ser., Atlas 214c, 40 p.

Cameron, C. C. (1970), *Peat deposits of northeastern Pennsylvania*, U.S. Geological Survey Bulletin 1317–A, 90 p.

Cooper, G. A. (1941), *New Devonian stratigraphic units*, Washington Academy of Science Journal, v. 31, p. 179–181.

Crowl, G. H. (1971), *Pleistocene geology and unconsolidated deposits of the Delaware Valley, Matamoras to Shawnee on Delaware, Pennsylvania*, Pennsylvania Geological Survey, 4th ser., General Geology Report 60, 40 p.

Crowl, G. H., and Sevon, W. D. (1980), *Glacial border deposits of late Wisconsinan age in northeastern Pennsylvania*, Pennsylvania Geological Survey, 4th ser., General Geology Report 71, 68 p.

Cunningham, R. L., Petersen, G. W., Matelski, R. P., and others (1972), *Laboratory characterization data and field descriptions of selected Pennsylvania soils*, Pennsylvania State University, College of Agriculture, Agricultural Experiment Station Agronomy Series 25, 805 p.

Davis, D. K. (1989), *Groundwater resources of Pike County, Pennsylvania*, Pennsylvania Geological Survey, 4th ser., Water Resource Report 65, 63 p.

Edgerton, C. D. (1969), *Peat bog investigations in northeastern Pennsylvania*, Pennsylvania Geological Survey, 4th ser., Information Circular 65, 53 p.

Epstein, A. G., Epstein, J. B., and Harris, L. D. (1977), *Conodont color alteration—an index to organic metamorphism*, U.S. Geological Survey Professional Paper 995, 27 p.

Epstein, J. B. (1969), *Surficial geology of the Stroudsburg quadrangle, Pennsylvania-New Jersey*, Pennsylvania Geological Survey, 4th ser., General Geology Report 57, 67 p.

Epstein, J. B., Sevon, W. D., and Glaeser, J. D. (1974), *Geology and mineral resources of the Lehighton and Palmerton quadrangles, Carbon and Northampton Counties, Pennsylvania*, Pennsylvania Geological Survey, 4th ser., Atlas 195cd, 460 p.

Faill, R. T. (1973), *Kink-band folding, Valley and Ridge province, Pennsylvania*, Geological Society of America Bulletin, v. 84, p. 1289–1314.

Fisher, D. W., Isachsen, Y. W., and Rickard, L. V., compilers and eds. (1970), *Geologic map of New York—Lower Hudson sheet,* New York State Museum and Science Service Map and Chart Series 15, scale 1:250,000.

Fletcher, F. W. (1963), *Regional stratigraphy of Middle and Upper Devonian non-marine rocks in southeastern New York*, in Shepps, V. C., ed., *Symposium on Middle and Upper Devonian stratigraphy of Pennsylvania and adjacent states*, Pennsylvania Geological Survey, 4th ser., General Geology Report 39, p. 25–42.

Fletcher, F. W., and Woodrow, D. L. (1970), *Geology and economic resources of the Pennsylvania portion of the Milford and Port Jervis 15-minute quadrangles*, Pennsylvania Geological Survey, 4th ser., Atlas 223, 64 p.

Fluhr, G. S. A. (1970), *Shohola 1600–1970: Highlights of history in northeastern Pennsylvania*, Shohola, Pennsylvania, E. S. Jarosz, 42 p.

Folk, R. L. (1954), *The distinction between grain size and mineral composition in sedimentary-rock nomenclature*, Journal of Geology, v. 62, no. 4, p. 344–359.

Goddard, E. N., Trask, P. D., DeFord, R. K., and others (1970), *Rock-color chart,* Boulder, Colo., Geological Society of America.

Hack, J. T., and Goodlett, J. C. (1960), *Geomorphology and forest ecology of a mountain region in the central Appalachians*, U.S. Geological Survey Professional Paper 347, 66 p.

Harms, J. C., Southard, J. B., Spearing, D. R., and Walker, R. G. (1975), *Depositional environments as interpreted from primary sedimentary structures and stratification sequences*, Society of Economic Paleontologists and Mineralogists Short Course 2, 161 p.

Kebblish, William, and Tuchman, R. J. (1983), *The mineral industry of Pennsylvania in 1981*, Pennsylvania Geological Survey, 4th ser., Information Circular 92, 16 p.

Krajewski, Stephen, and Williams, E. G. (1971), *Upper Devonian flagstones from northeastern Pennsylvania*, Pennsylvania State University, College of Earth and Mineral Sciences Special Publication 3–71, 185 p.

Levine, E. R. (1981), *Time as a factor in the genesis of soils formed from glacial till in northeastern Pennsylvania*, University Park, Pennsylvania State University, M.S. thesis, 122 p.

Levine, E. R., and Ciolkosz, E. J. (1983), *Soil development in till of various ages in northeastern Pennsylvania*, Quaternary Research, v. 19, no. 1, p. 85–99.

Lewis, H. C. (1884), *Report on the terminal moraine in Pennsylvania and western New York*, Pennsylvania Geological Survey, 2nd ser., Report Z, 299 p.

Matthews, A. (1886), *History of Wayne, Pike and Monroe Counties, Pennsylvania*, Philadelphia, R. T. Peck and Co., p. 969–971.

Mikutowicz, W. G., and Schenck, G. H. K. (1970a), *Marketing and distribution of Pennsylvania bluestone*, Pennsylvania State University, College of Earth and Mineral Sciences Special Publication 3–70, 73 p.

________ (1970b), *Future strategies for Pennsylvania bluestone industry*, Pennsylvania State University, College of Earth and Mineral Sciences Special Publication 4–70, 25 p.

Munsell Color (1975), *Munsell Soil Color Charts,* Baltimore, Md., Munsell Color.

Palmer, J. K., and MacNamara, E. E. (1972), *Some properties of a peat bog located in Pike County, Pennsylvania*, Pennsylvania Academy of Science Proceedings, v. 46, p. 117–121.

Prosser, C. S. (1894), *The Devonian System of eastern Pennsylvania and New York*, U.S. Geological Survey Bulletin 120, 81 p.

Rickard, L. V. (1975), *Correlation of the Silurian and Devonian rocks in New York State,* New York State Museum and Science Service Map and Chart Series 24, 16 p.

Rogers, H. D. (1858), *The geology of Pennsylvania—A government survey*, Philadelphia, J. B. Lippincott and Co., v. 1, 586 p.; v. 2, 1046 p.

Schultz, L. D. (1973), *Bedrock and surficial geology of the Pennsylvania portion of the Culvers Gap, Flatbrookville, and Lake Maskenozha quadrangles*, Pennsylvania Geological Survey, 4th ser., Open-File Report, 88 p.

________ (1974), *The stratigraphy of the Trimmers Rock Formation in northeastern Pennsylvania*, Bethlehem, Pa., Lehigh University, Ph.D. thesis, 132 p.

Sevon, W. D. (1972), *Late Wisconsinan periglacial boulder deposits in northeastern Pennsylvania* [abs.], Geological Society of America Abstracts with Programs, v. 4, no. 1, p. 43–44.

________ (1974), *Relative age and sequence of glacial deposits in Carbon and Monroe Counties, Pennsylvania* [abs.], Geological Society of America Abstracts with Programs, v. 6, no. 1, p. 71.

________ (1975a), *Geology and mineral resources of the Christmans and Pohopoco Mountain quadrangles, Carbon and Monroe Counties, Pennsylvania*, Pennsylvania Geological Survey, 4th ser., Atlas 195ab, scale 1:24,000, 2 sheets.

________ (1975b), *Geology and mineral resources of the Hickory Run and Blakeslee quadrangles, Carbon and Monroe Counties, Pennsylvania*, Pennsylvania Geological Survey, 4th ser., Atlas 194cd, scale 1:24,000, 2 sheets.

________ (1975c), *Geology and mineral resources of the Tobyhanna and Buck Hill Falls quadrangles, Monroe County, Pennsylvania*, Pennsylvania Geological Survey, 4th ser., Atlas 204ab, scale 1:24,000, 2 sheets.

Sevon, W. D. (1979), *Devonian sediment-dispersal systems in Pennsylvania* [abs.], Geological Society of America Abstracts with Programs, v. 11, no. 1, p. 53.

________ (1981), *The Middle and Upper Devonian clastic wedge in northeastern Pennsylvania*, in Enos, Paul, ed., *Guidebook for field trips in south-central New York*, Annual Meeting of New York State Geological Association, 53rd, Binghamton, N.Y., 1981, Guidebook, p. 31–64.

________ (1984), *A sandstone weathering rate from northeastern Pennsylvania* [abs.], Geological Society of America Abstracts with Programs, v. 16, no. 1, p. 63.

Sevon, W. D., and Berg, T. M. (1978), *Geology and mineral resources of the Skytop quadrangle, Monroe and Pike Counties, Pennsylvania*, Pennsylvania Geological Survey, 4th ser., Atlas 214a, 33 p.

________ (1979), *Pennsylvania shale-chip rubble*, Pennsylvania Geology, v. 10, no. 6, p. 2–7.

Sevon, W. D., Crowl, G. H., and Berg, T. M. (1975), *The late Wisconsinan drift border in northeastern Pennsylvania*, Annual Field Conference of Pennsylvania Geologists, 40th, Bartonsville, Pa., Guidebook, 108 p.

Sharpe, C. F. S. (1938), *Landslides and related phenomena—A study of mass-movements of soil and rock*, New York, Columbia University Press, 137 p.

Steel, R. J., and Aasheim, S. M. (1978), *Alluvial sand deposition in a rapidly subsiding basin (Devonian, Norway)*, in Miall, A. D., ed., *Fluvial sedimentology*, Canadian Society of Petroleum Geologists Memoir 5, p. 385–412.

Stone, R. W. (1923), *Flagstone industry in northeastern Pennsylvania*, Pennsylvania Geological Survey, 4th ser., Progress Report 72, 7 p.

________ (1932), *Building stones of Pennsylvania*, Pennsylvania Geological Survey, 4th ser., Mineral Resource Report 15, p. 254–258.

Taylor, D. C. (1969), *Soil survey of Pike County, Pennsylvania*, U.S. Department of Agriculture, Soil Conservation Service, 83 p.

Thoms, R. E., and Berg, T. M. (1985), *Interpretation of bivalve trace fossils in fluvial beds of the basal Catskill Formation (Late Devonian), eastern U.S.A.*, in Curran, H. A., ed., *Biogenic structures; their use in interpreting depositional environments,* Society of Economic Paleontologists and Mineralogists Special Publication 35, p. 13–20.

Thompson, A. M., and Sevon, W. D. (1982), *Excursion 19B: comparative sedimentology of Paleozoic clastic wedges in the central Appalachians, U.S.A.*, International Association of Sedimentologists, International Congress on Sedimentology, 11th, Hamilton, Ontario, Canada, Guidebook, 136 p.

U.S. Army Corps of Engineers (1967), *Tocks Island Reservoir, site geology, design memorandum #6*, USA Engineer District, Philadelphia, Pennsylvania.

U.S. Department of Commerce, Bureau of Census (1981), *1980 Census of population and housing: Pennsylvania*, Advance Reports PHC80–V–40, 46 p.

White, I. C. (1881), *The geology of Susquehanna County and Wayne County*, Pennsylvania Geological Survey, 2nd ser., Report G5, 243 p.

White, I. C. (1882), *The geology of Pike and Monroe Counties*, Pennsylvania Geological Survey, 2nd ser., Report G6, 407 p.

________ (1885), *The geology of Huntingdon County*, Pennsylvania Geological Survey, 2nd ser., Report T3, 471 p.

Willard, Bradford (1936), *Continental Upper Devonian of northeastern Pennsylvania*, Geological Society of America Bulletin, v. 47, p. 565–607.

________ (1938), *A Paleozoic section at Delaware Water Gap*, Pennsylvania Geological Survey, 4th ser., General Geology Report 11, 35 p.

Willard, Bradford, Swartz, F. M., and Cleaves, A. B. (1939), *The Devonian of Pennsylvania*, Pennsylvania Geological Survey, 4th ser., General Geology Report 19, 481 p.

Wood, G. H., Jr. (1974a), *Geologic map of the Nesquehoning quadrangle, Carbon and Schuylkill Counties, Pennsylvania*, U.S. Geological Survey Geologic Quadrangle Map GQ–1132, scale 1:24,000.

________ (1974b), *Geologic map of the Tamaqua quadrangle, Carbon and Schuylkill Counties, Pennsylvania*, U.S. Geological Survey Geologic Quadrangle Map GQ–1133, scale 1:24,000.

Woodrow, D. L., and Sevon, W. D., eds. (1985), *The Catskill delta*, Geological Society of America Special Paper 201, 246 p.

# GLOSSARY

*Ablation.* All processes by which snow or ice is lost from the surface of a glacier, including melting, evaporation (sublimation), and wind erosion.

*Alluvial.* A term referring to clay, silt, sand, and gravel deposited by a stream or running water.

*Alluvial plain.* A gently sloping land surface produced by deposition of sediment by rivers.

*Authigenic.* Formed or generated in place.

*Amplitude.* The height of a fold.

*Anticline.* A fold in layered rocks in which the layers dip away from the hinge; it is convex upward.

*Bed.* A sedimentary layer of relatively homogeneous material separated from adjacent layers by well-defined boundary surfaces.

*Bedrock.* A general term for solid rock that underlies soil or other surficial material.

*Bentonite.* A dense, light-colored rock composed of clay minerals that formed by alteration of a volcanic ash bed; it contains feldspars and zircons.

*Boulder.* A rock having a diameter greater than 256 mm; larger than a cobble.

*Bouma cycle.* A fixed succession of five intervals that make up a complete sequence of turbidite.

*Brachiopod.* A marine animal that is similar to a clam, but whose two shells are unequal in size.

*Braided river.* A river that divides into or follows an interlacing network of several small, shallow channels separated from each other by islands or bars.

*Bryozoan.* A marine animal that has a calcareous skeleton and usually lives on a shallow sea floor in a colony.

*Calcareous.* Pertaining to rocks containing calcite, in particular rocks in which the grains are cemented with calcite.

*Channel.* A watercourse eroded from underlying sediments of an alluvial floodplain and which may be filled with similar sediments.

*Channery.* Pertaining to soils that contain thin, flat fragments of sandstone, limestone, or schist, as much as 6 inches in length.

*Chloritization.* The replacement of clay minerals such as illite by chlorite.

*Chroma.* The quality of a color that pertains to its vividness.

*Clast.* An individual grain or fragment of a sediment or rock.

*Clay.* A rock or mineral fragment smaller than very fine silt (less than 0.0039 mm in size); a deposit containing mostly clay-sized particles.

*Cleavage.* Closely spaced planar surfaces produced by deformation, along which a rock tends to split or break.

*Coarse grained.* See *Grain size.*

*Colluvium.* Loose, incoherent mass of soil and rock material transported and deposited by mass wasting.

*Columnal.* A single, commonly circular plate that was a segment of a column (stem) supporting a marine animal, such as a crinoid.

*Concretion.* A rounded, commonly ellipsoidal or irregularly shaped, mineral mass precipitated from aqueous solution in the pores or cavities of a sedimentary rock.

*Conglomeratic.* Pertaining to a clastic sedimentary rock containing significant amounts of gravel.

*Conodont.* A small, toothlike or platelike structure, generally 0.3 to 1.5 mm in length, composed of laminae of crystalline apatite; it is of biologic origin, but unknown affinity.

*Contact.* A surface between two different rock units.

*Crinoid.* A marine animal having a globular body and radiating "arms" and connected to the substrate by a column (stem).

*Crossbeds.* Minor beds within a sedimentary layer that are inclined to the surface of the layer.

*Delta.* A tract of land, part subaerial and part subaqueous, formed by sediment deposition near the mouth of a river.

*Devonian.* The geologic time period between 405 and 365 million years ago; pertaining to any deposit formed during this time period.

*Dip.* The vertical angle between a surface, such as bedding or a fault, and the horizontal, measured perpendicular to the strike of the surface.

*Drift.* All rock material transported by a glacier and deposited directly or indirectly by the ice, or by running water issuing from a glacier.

*Fault.* A surface or zone of rock fracture along which there has been displacement.

*Fine grained.* See *Grain size.*

*Fissile; fissility.* The capability of being easily split along closely spaced planes; the property of a rock (usually a shale) that causes it to split easily, parallel to bedding.

*Floodplain.* A relatively flat terrain in a valley adjacent to a stream or river channel, which is covered with water when the stream overflows its banks.

*Fluvial.* Pertaining to a river or stream.

*Fold.* A curve or bend of an initially planar feature such as a bedding plane; it is usually a result of deformation.

*Fold axis.* A line about which the fold layers are curved or bent.

*Fold wavelength.* The distance between the hinges of adjacent anticlines (or synclines).

*Formation.* A mappable body of rock, commonly in tabular form and of considerable thickness, containing a distinctive lithology or combination of lithologies.

*Fossil; fossiliferous.* The remains, trace, or imprint of a plant or animal that has been preserved in the rock; pertaining to a rock containing fossils.

*Fracture.* A break in a rock induced by stress.

*Fracture trace.* A natural linear feature, less than a mile in length, consisting of topographic, vegetation, or soil tonal alignments visible principally on aerial photographs.

*Frost riving.* The mechanical disintegration of a rock due to great pressure exerted by freezing water contained in cracks or pores in the rock.

*Glacial till.* A deposit produced by a glacier or ice sheet.

*Glacial striation.* A scratch or furrow inscribed on a rock surface by rock debris within glacial ice. Scratches on in-situ bedrock surfaces parallel the direction of ice flow.

*Glacier.* A large mass of ice formed on land by the compaction of snow; it moves downslope or across the land by creep and flow.

*Glaciofluvial.* Pertaining to meltwater streams flowing from glaciers or to the deposits formed by such streams.

*Grain.* A mineral or rock particle generally less than a few millimeters in diameter.

*Grain size.* A term pertaining to the mean size of the constituent particles of a rock. The terms and their size limits are as follows:

| *Grain size* | *Upper size limits (mm)* |
|---|---|
| Boulder | >256 |
| Cobble | 256 |
| Pebble | 64 |
| Granule | 4 |
| Very coarse sand | 2 |
| Coarse sand | 1 |
| Medium sand | .5 |
| Fine sand | .25 |
| Very fine sand | .125 |
| Silt | .0625 |
| Clay | .0039 |

*Gravel.* An unconsolidated deposit, consisting predominantly of pebbles (particles larger than sand; greater than 2 mm in diameter).

*Holocene.* The portion of the Quaternary Period from 10,000 years ago to the present; pertaining to any deposit formed during this period.

*Homocline.* General term for a structural unit in which the strata have the same dip.

*Interbed.* A bed of one lithology between beds of a different lithology.

*Intraformational conglomerate.* A conglomerate in which the clasts are essentially contemporaneous in origin with the matrix and were derived from nearby, partly consolidated, similar sediment.

*Joint.* A planar fracture having little or no visible displacement.

*Kink band.* A tabular, more or less parallel-sided rock zone within which bedding has been rotated relative to bedding outside the zone.

*Layer.* A tabular body of rock; a bed.

*Limb.* One side of a fold, within which bedding normally dips in the same direction.

*Lithology.* The physical character of a rock, including color, composition, grain size, and so forth.

*Load cast.* A bulb-shaped protuberance on the bottom of a sandstone or siltstone bed produced by unequal settling on a soft substratum.

*Meandering river.* A river having a pattern of successive sharp, sinuous curves.

*Medium grained.* See *Grain size.*

*Member.* A portion, or subdivision, of a formation in which the rocks are different in some way from the rocks of the rest of the formation.

*Microcrystalline.* The constituents of a rock that are visible only with a microscope.

*Microscopic petrology.* The determination of the mineral composition of a rock by use of rock thin sections and a polarizing microscope.

*Olean.* A term used to name glacial drift of late Wisconsinan (Woodfordian) age.

*Ordovician.* The geologic time period between 500 and 430 million years ago; pertaining to any deposit formed during this period of time.

*Orogeny.* A period of mountain building; also, the process by which the structures (folds, faults, etc.) within mountain areas are formed.

*Outcrop.* Rock exposed at the surface of the earth.

*Outwash.* Sand and/or gravel removed from a glacier by water and deposited near its margins.

*Parting.* A surface in a rock mass along which the rock separates easily (e.g., a bedding surface or a fracture surface).

*Parting lineation.* A primary sedimentary structure consisting of poorly defined parallel grooves on a bedding surface. The grooves parallel the direction of current flow at the time of deposition.

*Parting-step lineation.* A primary sedimentary structure consisting of parallel to subparallel, steplike ridges where the parting between bedding surfaces cuts across several adjacent laminae. The ridges parallel the direction of current flow at the time of deposition.

*Pelecypod.* A marine animal that has two equal shells; a clam.

*Periglacial.* Pertaining to the region or environment adjacent to a glacier.

*Pleistocene.* The portion of the Quaternary Period from 2 million to 10,000 years ago (the "Ice Age"); pertaining to any deposit formed during this period.

*Plunge.* The vertical angle between the horizontal and a linear structure, such as a fold axis or a slickenline.

*Prodelta.* The part of a delta complex that is below the effective depth of wave erosion and beyond the delta front.

*Quadrangle.* A four-sided area bounded by latitudinal parallels and longitudinal meridians, commonly represented by topographic and geologic maps.

*Quartz.* A generally colorless silicon oxide; a very common rock-forming mineral.

*Quartzite.* A very hard sandstone consisting almost entirely of quartz grains cemented with silica.

*Quaternary.* The time period from 2 million years ago to the present; pertaining to any deposit formed during this time period.

*Ripple mark.* An undulatory sedimentary surface consisting of small subparallel ridges and troughs, and commonly found in sandstone and siltstone.

*Rock.* Any naturally formed material composed of one or more minerals.

*Sand.* An unconsolidated material consisting predominantly of sand-sized particles (between 0.0625 and 2 mm in diameter).

*Sandstone.* A clastic sedimentary rock composed predominantly of sand-sized particles.

*Sedimentary rock.* A rock resulting from the cementation or consolidation of loose sediment, or chemical precipitation from a solution (such as seawater).

*Sediment-dispersal system.* A major river that deposits sediment in the same general area for a long period of time.

*Shale.* A fine-grained sedimentary rock consisting predominantly of clay-sized particles (less than 0.0039 mm in diameter) and possessing a bed-parallel fissility.

*Silt.* An unconsolidated material consisting predominantly of silt-sized particles (between 0.0039 and 0.0625 mm in diameter).

*Siltstone.* A clastic sedimentary rock composed predominantly of silt-sized particles.

*Silurian.* The geologic time period between 430 and 405 million years ago; pertaining to any deposit formed during this time period.

*Soil.* An unconsolidated material near the surface of the earth in which vegetation grows and identifiable weathering-produced zonation occurs.

*Stratified drift.* Sorted and layered material deposited by a glacial meltwater stream.

*Strike.* The trend of a horizontal line on a surface, such as a bedding or a fault surface.

*Surficial.* Pertaining to unconsolidated material overlying bedrock.

*Thin section.* A fragment of rock that has been ground to a thickness of 0.03 mm, polished, and mounted between pieces of glass for viewing with a polarizing microscope.

*Turbidite.* A sediment or rock deposited from a bottom-flowing current laden with suspended sediment; it is generally characterized by graded bedding and a distinctive vertical sequence of sedimentary structures. Proximal turbidites are deposited near the source area of the sediment, whereas distal turbidites are deposited far out in the sedimentary basin.

*Weathered.* A term applied to a rock that has been altered from its original state by chemical and mechanical processes at or near the surface of the earth.

*Wisconsinan.* The last of the four major Pleistocene continental glaciations in North America; it lasted from 70,000 to 10,000 years ago.

*Woodfordian.* A late phase of the Wisconsinan continental glaciation; it lasted from 25,000 to 12,000 years ago.

# FACTORS FOR CONVERTING INCH-POUND UNITS TO INTERNATIONAL SYSTEM UNITS (SI)

| *Multiply inch-pound units* | *by* | *To obtain SI units* |
|---|---|---|
| | *Length* | |
| inch (in.) | 25.40 | millimeter (mm) |
| inch (in.) | 2.540 | centimeter (cm) |
| foot (ft) | .3048 | meter (m) |
| mile (mi) | 1.609 | kilometer (km) |
| | *Area* | |
| acre | .4047 | hectare |
| square mile ($mi^2$) | 2.590 | square kilometer ($km^2$) |
| | *Volume* | |
| cubic foot ($ft^3$) | .02832 | cubic meter ($m^3$) |
| cubic yard ($yd^3$) | .7646 | cubic meter ($m^3$) |
| | *Mass* | |
| pound, avoirdupois (lb) | 453.6 | gram (g) |
| ton, short | .9072 | megagram (Mg) |
| | *Density* | |
| pound per cubic foot ($lb/ft^3$) | 16.02 | kilogram per cubic meter ($kg/m^3$) |
| pound per cubic yard ($lb/yd^3$) | .5933 | kilogram per cubic meter ($kg/m^3$) |
| | *Temperature* | |
| degree Fahrenheit (°F) | °C = 5/9(°F − 32) | degree Celsius (°C) |

# APPENDICES

## APPENDIX 1. RELATIVE AMOUNTS OF GRAVEL AND SAND-SILT-CLAY (SSC) IN 81 SAMPLES OF OLEAN TILL FROM PIKE COUNTY AND NEIGHBORING AREAS[1, 2]

### Area A

| Sample number | SSC (percent) | Gravel (percent) |
|---|---|---|
| 1 | 82 | 18 |
| 2 | 73 | 27 |
| 3 | 75 | 25 |
| 4 | 80 | 20 |
| 5 | 71 | 29 |
| 6 | 71 | 29 |
| 7 | 70 | 30 |
| 8 | 68 | 32 |
| 9 | 82 | 18 |
| 10 | 81 | 19 |
| 11 | 61 | 39 |
| 12 | 52 | 48 |
| 13 | 81 | 19 |
| 14 | 77 | 23 |
| 15 | 62 | 38 |
| 16 | 74 | 26 |
| 17 | 60 | 40 |
| 18 | 78 | 22 |
| 19 | 70 | 30 |
| 20 | 80 | 20 |
| 21 | 64 | 36 |
| 22 | 76 | 24 |
| 23 | 51 | 49 |
| 24 | 93 | 7 |
| 25 | 75 | 25 |
| 26 | 81 | 19 |
| 27 | 59 | 41 |
| 28 | 76 | 24 |
| 29 | 57 | 43 |
| 30 | 56 | 44 |
| 31 | 54 | 46 |
| 32 | 83 | 17 |
| 33 | 71 | 29 |
| 34 | 84 | 16 |
| 35 | 65 | 35 |
| 36 | 50 | 50 |
| 37 | 61 | 39 |
| 38 | 76 | 24 |

### Area A *(Continued)*

| Sample number | SSC (percent) | Gravel (percent) |
|---|---|---|
| 39 | 81 | 19 |
| 40 | 74 | 26 |
| 41 | 74 | 26 |
| 46 | 55 | 45 |
| 47 | 72 | 28 |
| 48 | 81 | 19 |
| 49 | 73 | 27 |
| 50 | 67 | 33 |
| 51 | 73 | 27 |
| 52 | 40 | 60 |
| 53 | 81 | 19 |
| 54 | 67 | 33 |
| 55 | 56 | 44 |
| 56 | 74 | 26 |
| 57 | 62 | 38 |
| 61 | 66 | 34 |
| 62 | 68 | 32 |
| 63 | 66 | 34 |
| 64 | 40 | 60 |
| 68 | 73 | 27 |
| 76 | 70 | 30 |
| 77 | 60 | 40 |
| 78 | 73 | 27 |
| 79 | 61 | 39 |
| 80 | 68 | 32 |
| 81 | 71 | 29 |
| Mean | 69 | 31 |

### Area B

| Sample number | SSC (percent) | Gravel (percent) |
|---|---|---|
| 42 | 56 | 44 |
| 44 | 71 | 29 |
| 58 | 49 | 51 |
| 70 | 63 | 37 |
| 71 | 63 | 37 |
| 75 | 56 | 44 |
| Mean | 60 | 40 |

### Area C

| Sample number | SSC (percent) | Gravel (percent) |
|---|---|---|
| 43 | 90 | 10 |
| 45 | 87 | 13 |
| 59 | 69 | 31 |
| 60 | 48 | 52 |
| 65 | 56 | 44 |
| 66 | 81 | 19 |
| 67 | 53 | 47 |
| 69 | 66 | 34 |
| 72 | 54 | 46 |
| 73 | 51 | 49 |
| 74 | 66 | 34 |
| Mean | 66 | 34 |
| Grand mean | 68 | 32 |

[1]Sample locations are shown in Figure 50.
[2]Values were obtained by weight measurement.

## APPENDIX 2. RELATIVE AMOUNTS OF SAND, SILT, AND CLAY IN 81 SAMPLES OF OLEAN TILL FROM PIKE COUNTY AND NEIGHBORING AREAS[1, 2]

### Area A

| Sample number | Sand (percent) | Silt (percent) | Clay (percent) |
|---|---|---|---|
| 1 | 46 | 18 | 36 |
| 2 | 52 | 25 | 23 |
| 3 | 46 | 26 | 28 |
| 4 | 43 | 38 | 19 |
| 5 | 48 | 26 | 26 |
| 6 | 48 | 22 | 30 |
| 7 | 50 | 24 | 26 |
| 8 | 54 | 24 | 22 |
| 9 | 54 | 26 | 20 |
| 10 | 50 | 30 | 20 |
| 11 | 60 | 25 | 15 |
| 12 | 50 | 30 | 20 |
| 13 | 49 | 30 | 21 |
| 14 | 50 | 24 | 26 |
| 15 | 54 | 28 | 18 |
| 16 | 58 | 27 | 15 |
| 17 | 65 | 21 | 14 |

### Area A (Continued)

| Sample number | Sand (percent) | Silt (percent) | Clay (percent) |
|---|---|---|---|
| 18 | 52 | 31 | 17 |
| 19 | 55 | 29 | 16 |
| 20 | 49 | 27 | 24 |
| 21 | 43 | 38 | 19 |
| 22 | 47 | 34 | 19 |
| 23 | 48 | 31 | 21 |
| 24 | 44 | 29 | 27 |
| 25 | 63 | 27 | 20 |
| 26 | 58 | 25 | 17 |
| 27 | 56 | 26 | 18 |
| 28 | 38 | 40 | 22 |
| 29 | 58 | 26 | 16 |
| 30 | 50 | 32 | 18 |
| 31 | 58 | 26 | 16 |
| 32 | 48 | 28 | 24 |
| 33 | 48 | 27 | 25 |
| 34 | 46 | 29 | 25 |
| 35 | 60 | 26 | 14 |
| 36 | 59 | 30 | 11 |
| 37 | 58 | 26 | 16 |
| 38 | 51 | 24 | 25 |
| 39 | 61 | 25 | 14 |
| 40 | 51 | 30 | 19 |
| 41 | 52 | 28 | 20 |
| 46 | 50 | 25 | 25 |
| 47 | 56 | 26 | 18 |
| 48 | 50 | 28 | 22 |
| 49 | 56 | 26 | 18 |
| 50 | 46 | 30 | 24 |
| 51 | 58 | 30 | 12 |
| 52 | 60 | 31 | 9 |
| 53 | 60 | 24 | 16 |
| 54 | 47 | 35 | 18 |
| 55 | 60 | 28 | 12 |
| 56 | 60 | 22 | 18 |
| 57 | 53 | 32 | 15 |
| 61 | 42 | 27 | 31 |
| 62 | 54 | 26 | 20 |
| 63 | 53 | 27 | 20 |
| 64 | 40 | 37 | 23 |
| 68 | 54 | 29 | 17 |
| 76 | 50 | 26 | 24 |
| 77 | 48 | 32 | 20 |
| 78 | 49 | 28 | 23 |
| 79 | 50 | 34 | 16 |
| 80 | 55 | 24 | 21 |
| 81 | 46 | 36 | 18 |
| Mean | 52 | 28 | 20 |

### Area B

| Sample number | Sand (percent) | Silt (percent) | Clay (percent) |
|---|---|---|---|
| 42 | 58 | 28 | 14 |
| 44 | 48 | 36 | 16 |
| 58 | 54 | 28 | 18 |
| 70 | 56 | 29 | 15 |
| 71 | 43 | 41 | 16 |
| 75 | 67 | 21 | 12 |
| Mean | 54 | 31 | 15 |

### Area C

| Sample number | Sand (percent) | Silt (percent) | Clay (percent) |
|---|---|---|---|
| 43 | 33 | 45 | 22 |
| 45 | 46 | 32 | 22 |

### Area C *(Continued)*

| Sample number | Sand (percent) | Silt (percent) | Clay (percent) |
|---|---|---|---|
| 59 | 44 | 34 | 22 |
| 60 | 48 | 35 | 17 |
| 65 | 48 | 32 | 20 |
| 66 | 40 | 37 | 23 |
| 67 | 46 | 32 | 22 |
| 69 | 50 | 30 | 20 |
| 72 | 43 | 40 | 17 |
| 73 | 47 | 35 | 18 |
| 74 | 50 | 34 | 16 |
| Mean | 45 | 35 | 20 |
| Grand mean | 51 | 29 | 20 |

[1]Sample locations are shown in Figure 50.
[2]Values were obtained by the hydrometer method of Bouyoucos (1928).

## APPENDIX 3. DESCRIBED SECTIONS

### Section 1. Reference section of Olean Till in Dingman Township, Pike County

The site is located in an old borrow area 0.1 mile northwest of the intersection of Overbrook Run Road and Cayuga Trail Road in Conasbaugh Lakes development at 41°18′45″N/74°59′27″W. The borrow area is on the southwest side of Overbrook Run Road, and the site is on the north side of the borrow area. Data from size analyses and pebble lithology identifications are given in percent.

| Depth (cm) | Soil horizon | Description[1] |
|---|---|---|
| 0–0.5 | $A_0$ | Organic matter. |
| 0.5–1.0 | $A_1$ | Sand and organic matter; dark brown (7.5YR3/2). |
| 1.0–10.5 | $A_2$ | Silt, fine sand, and pebbles; massive; strong brown (7.5YR5/6). Pebbles: sandstone, 92; siltstone, 8. |
| 10.5–27.5 | $A_3$ | Silt, sand, and pebbles; platy; brown (7.5YR5/4). Size: sand, 48; silt, 30; clay, 22. Pebbles: gray sandstone, 85; gray siltstone, 4; red siltstone, 8; white quartz, 3. |
| 27.5–45.0 | B | Sand, silt, clay, and pebbles; porous; abundant clay skins; transitional from platy above to blocky below; reddish brown (5YR5/3). Size: sand, 52; silt, 28; clay, 20. Pebbles: gray sandstone, 86; red sandstone, 5; gray siltstone, 7; red siltstone, 1. |
| 45–>200 | C | Sand, silt, clay, and pebbles; blocky; stony; reddish brown (5YR5/3). Size at 70 cm: sand, 45; silt, 25; clay, 20. Pebbles at 70 cm: gray sandstone, 58; red sandstone, 3; gray siltstone, 11; red siltstone, 14; gray claystone, 9; red claystone, 3; white quartz, 3. Size at 200 cm: sand, 48; silt, 27; clay, 25. Pebbles at 200 cm: gray sandstone, 76; gray siltstone, 8; red siltstone, 10; red claystone, 5. |

## Section 2. Reference section of Olean Till in Lehman Township, Pike County

The site is located in a landslide face on the northeast side of a secondary road about 1.1 miles northeast of Bushkill and 0.2 mile northwest of U.S. Route 209 at 41°06′32″N/74°59′16″W. Data from size analyses and pebble lithology identifications are given in percent.

| Depth (cm) | Soil horizon | Description[1] |
|---|---|---|
| 0–2 | $A_1$ | Organic matter containing some sand and pebbles. |
| 2–20 | $A_2$ | Sand, silt, clay, and pebbles; crumbly; colluviated zone having pebbles of angular fragments from the Mahantango Formation; yellowish brown (10YR5/8). Size: sand, 55; silt, 31; clay, 14. Pebbles: gray sandstone, 1; gray siltstone, 9; gray claystone, 89; white quartz, 1. |
| 20–39 | $B_c$ | Sand, silt, clay, and pebbles; crumbly; colluviated zone containing coarser grained pebbles than $A_2$; yellowish brown (10YR5/4). Size: sand, 44; silt, 36; clay, 20. Pebbles: gray sandstone, 8; gray siltstone, 92. |
| 39–63 | $B_t$ | Sand, silt, clay, and pebbles; unsorted; slightly platy; abundant clay skins in near fretwork mass; sharp and wavy upper contact; gradational lower contact; yellowish brown (10YR5/4). Size: sand, 42; silt, 40; clay, 18. Pebbles: gray sandstone, 2; gray siltstone, 96; gray claystone, 1; white quartz, 1. |
| 65–563 | C | Sand, silt, clay, and pebbles; unsorted; compact; oxidized; prismatic to platy structure; weathered along prism faces to at least 200 cm; brown (10YR5/3). Size at 115 cm: sand, 55; silt, 33; clay, 12. Pebbles at 115 cm: gray sandstone, 3; gray siltstone, 97. Size at 375 cm: sand, 31; silt, 43; clay, 26. Pebbles at 375 cm: gray sandstone, 9; gray siltstone, 91. |
| 563–900 | | Covered. Scattered small exposures along face indicate sharp, irregular base of oxidized zone having relief of 100 to 200 cm. |
| >900 | D | Sand, silt, clay, and pebbles; unsorted; very compact; structureless; unoxidized; slightly calcareous; dark gray (7.5YR4/0). Because of nature of exposure, true depth of oxidized zone may be less than indicated here. Size: sand, 62; silt, 26; clay, 12. Pebbles: gray conglomeratic sandstone, 1; gray sandstone, 10; red sandstone, 1; gray siltstone, 88; white quartz, 1. |

[1]Colors are based on the *Munsell Soil Color Charts* (Munsell Color, 1975).

# APPENDIX 4. PRELIMINARY CLAY-SHALE EVALUATIONS[1]

## Sample 212–C–1

Unit: Catskill Formation, Lackawaxen Member
Quadrangle: White Mills
Latitude/longitude: 41°30′22″N/75°08′31″W
Sampled interval: 12 feet

RAW PROPERTIES:
Water of plasticity (percent): 13.1
Color: Brown
pH: 7.6
HCl effervescence: None

Working properties: Short
Drying shrinkage (percent): 0.0
Dry strength: Fair
Preliminary bloating test: Negative

SLOW-FIRING TEST:

| Temperature (°C) | Color[2] | Hardness (Mohs scale) | Percent linear shrinkage | Percent absorption | Percent apparent porosity | Bulk density (g/cm$^3$) |
|---|---|---|---|---|---|---|
| 1000 | 2.5YR6/8 | 3 | 2.5 | 12.1 | 24.5 | 2.03 |
| 1050 | 2.5YR5/8 | 3 | 2.5 | 9.7 | 20.5 | 2.11 |
| 1100 | 2.5YR5/6 | 4 | 5.0 | 6.5 | 14.5 | 2.25 |
| 1150 | 2.5YR4/4 | 5 | 7.5 | 1.0 | 2.4 | 2.43 |
| 1200 | — | — | Melted | — | — | — |
| 1250 | — | — | — | — | — | — |

Potential use: Structural clay products (e.g., building brick, floor brick) at 1100 to 1150°C.

## Sample 213–A–1

Unit: Catskill Formation, Lackawaxen Member
Quadrangle: Hawley
Latitude/longitude: 41°24′33″N/75°10′59″W
Sampled interval: 16 feet

RAW PROPERTIES:
Water of plasticity (percent): 12.3
Color: Brown
pH: 7.6
HCl effervescence: None

Working properties: Short
Drying shrinkage (percent): 0.0
Dry strength: Poor
Preliminary bloating test: Negative

SLOW-FIRING TEST:

| Temperature (°C) | Color[2] | Hardness (Mohs scale) | Percent linear shrinkage | Percent absorption | Percent apparent porosity | Bulk density (g/cm$^3$) |
|---|---|---|---|---|---|---|
| 1000 | 2.5YR6/8 | 3 | 0.0 | 12.7 | 25.6 | 2.01 |
| 1050 | 2.5YR6/6 | 3 | 2.5 | 9.0 | 19.2 | 2.12 |
| 1100 | 2.5YR5/6 | 4 | 5.0 | 6.6 | 14.7 | 2.22 |
| 1150 | 2.5YR4/4 | 5 | 5.0 | 1.9 | 4.5 | 2.36 |
| 1200 | — | — | Melted | — | — | — |
| 1250 | — | — | — | — | — | — |

Potential use: Structural clay products (e.g., building brick, floor brick) at 1100 to 1150°C.

[1]Based on data obtained from the Tuscaloosa Metallurgy Research Laboratory of the U.S. Bureau of Mines. Data are preliminary in nature and will not suffice for plant or process design.
[2]Colors are based on the Munsell system (American Society of Testing and Materials, 1975).

## Sample 213–C–1

Unit: Catskill Formation, Poplar Gap and Packerton Members, undivided
Quadrangle: Promised Land
Latitude/longitude: 41°21′45″N/75°09′38″W
Sampled interval: 20 feet

RAW PROPERTIES:
Water of plasticity (percent): 14
Color: Brown
pH: 7.7
HCl effervescence: Slight
Working properties: Short
Drying shrinkage (percent): 0.0
Dry strength: Poor
Preliminary bloating test: Negative

SLOW-FIRING TEST:

| Temperature (°C) | Color[2] | Hardness (Mohs scale) | Percent linear shrinkage | Percent absorption | Percent apparent porosity | Bulk density (g/cm$^3$) |
|---|---|---|---|---|---|---|
| 1000 | 2.5YR6/6 | 3 | 0.0 | 13.9 | 27.3 | 1.96 |
| 1050 | 2.5YR6/6 | 3 | 2.5 | 10.3 | 21.3 | 2.07 |
| 1100 | 2.5YR5/6 | 3 | 5.0 | 8.5 | 18.2 | 2.14 |
| 1150 | 2.5YR4/4 | 4 | 5.0 | 4.8 | 10.8 | 2.24 |
| 1200 | — | — | Melted | — | — | — |
| 1250 | — | — | — | — | — | — |

Potential use: Marginal for structural clay products (e.g., building brick) at 1150°C.

Remarks: Short firing range. Presence of carbonates could cause problems.

## Sample 213–C–2

Unit: Catskill Formation, Lackawaxen Member
Quadrangle: Promised Land
Latitude/longitude: 41°21′54″N/75°09′08″W
Sampled interval: 15 feet

RAW PROPERTIES:
Water of plasticity (percent): 12.3
Color: Brown
pH: 7.8
HCl effervescence: Slight
Working properties: Short
Drying shrinkage (percent): 2.5
Dry strength: Poor
Preliminary bloating test: Negative

SLOW-FIRING TEST:

| Temperature (°C) | Color[2] | Hardness (Mohs scale) | Percent linear shrinkage | Percent absorption | Percent apparent porosity | Bulk density (g/cm$^3$) |
|---|---|---|---|---|---|---|
| 1000 | 2.5YR6/8 | 3 | 2.5 | 11.7 | 23.7 | 2.03 |
| 1050 | 2.5YR6/6 | 3 | 5.0 | 9.4 | 20.0 | 2.13 |
| 1100 | 2.5YR5/6 | 4 | 5.0 | 6.0 | 13.4 | 2.24 |
| 1150 | 2.5YR4/4 | 4 | 5.0 | 2.0 | 4.8 | 2.38 |
| 1200 | — | — | Melted | — | — | — |
| 1250 | — | — | — | — | — | — |

Potential use: Structural clay products (e.g., building brick, floor brick) at 1100 to 1150°C.

Remarks: Presence of carbonates could cause problems.

## Sample 214-A-1

Unit: Catskill Formation, Delaware River Member
Quadrangle: Skytop
Latitude/longitude: 41°14′58″N/75°14′22″W
Sampled interval: 20 feet

RAW PROPERTIES:
Water of plasticity (percent): 20.3
Color: Brown
pH: 6.9
HCl effervescence: None

Working properties: Plastic
Drying shrinkage (percent): 2.5
Dry strength: Good
Preliminary bloating test: Negative

SLOW-FIRING TEST:

| Temperature (°C) | Color[2] | Hardness (Mohs scale) | Percent linear shrinkage | Percent absorption | Percent apparent porosity | Bulk density (g/cm$^3$) |
|---|---|---|---|---|---|---|
| 1000 | 2.5YR6/6 | 2 | 2.5 | 16.1 | 29.9 | 1.86 |
| 1050 | 2.5YR6/6 | 3 | 2.5 | 13.5 | 26.1 | 1.94 |
| 1100 | 2.5YR5/6 | 3 | 2.5 | 10.7 | 21.8 | 2.04 |
| 1150 | 10R4/4 | 4 | 5.0 | 7.3 | 15.8 | 2.17 |
| 1200 | 10R4/2 | 4 | 7.5 | 3.1 | 7.1 | 2.28 |
| 1250 | — | — | Melted | — | — | — |

Potential use: Structural clay products (e.g., building brick, floor brick) at 1150 to 1200°C.

## Sample 214-D-1

Unit: Mahantango Formation
Quadrangle: Bushkill
Latitude/longitude: 41°05′11″N/75°01′24″W
Sampled interval: 24 feet

RAW PROPERTIES:
Water of plasticity (percent): 13
Color: Gray
pH: 7.4
HCl effervescence: None

Working properties: Short
Drying shrinkage (percent): 0.0
Dry strength: Poor
Preliminary bloating test: Positive

SLOW-FIRING TEST:

| Temperature (°C) | Color[2] | Hardness (Mohs scale) | Percent linear shrinkage | Percent absorption | Percent apparent porosity | Bulk density (g/cm$^3$) |
|---|---|---|---|---|---|---|
| 1000 | 2.5YR6/8 | 3 | 2.5 | 13.1 | 25.8 | 1.97 |
| 1050 | 2.5YR5/8 | 3 | 2.5 | 10.7 | 22.1 | 2.06 |
| 1100 | 2.5YR4/6 | 4 | 5.0 | 6.9 | 15.2 | 2.21 |
| 1150 | 2.5YR3/4 | 4 | 5.0 | 2.9 | 6.6 | 2.26 |
| 1200 | — | — | Expanded | — | — | — |
| 1250 | — | — | — | — | — | — |

Sample 214-D-1 *(Continued)*

PRELIMINARY BLOATING TEST:

| Temperature (°C) | Percent absorption | Bulk density (g/cm³) | (lb/ft³) | Remarks |
|---|---|---|---|---|
| 1050 | 18.9 | 1.73 | 107.7 | Slight expansion. |
| 1100 | 17.6 | 1.63 | 101.9 | do. |
| 1150 | 13.9 | .91 | 56.5 | Good pore structure. |
| 1200 | 13.1 | .67 | 46.2 | do. |

Potential use: Structural clay products (e.g., building brick, floor brick) at 1100 to 1150°C. Lightweight aggregate at 1150 to 1200°C.

## Sample 214-D-2

Unit: Mahantango Formation
Quadrangle: Bushkill
Latitude/longitude: 41°05′11″N/75°01′24″W
Sampled interval: 30 feet

RAW PROPERTIES:
Water of plasticity (percent): 13.2
Color: Gray
pH: 7.8
HCl effervescence: None

Working properties: Short
Drying shrinkage (percent): 2.5
Dry strength: Poor
Preliminary bloating test: Positive

SLOW-FIRING TEST:

| Temperature (°C) | Color[2] | Hardness (Mohs scale) | Percent linear shrinkage | Percent absorption | Percent apparent porosity | Bulk density (g/cm³) |
|---|---|---|---|---|---|---|
| 1000 | 2.5YR6/8 | 3 | 2.5 | 12.5 | 25.0 | 2.00 |
| 1050 | 2.5YR5/6 | 3 | 2.5 | 9.9 | 20.7 | 2.09 |
| 1100 | 2.5YR4/6 | 4 | 5.0 | 6.1 | 13.7 | 2.22 |
| 1150 | 10R4/4 | 5 | 5.0 | 2.6 | 5.8 | 2.24 |
| 1200 | — | — | Expanded | — | — | — |
| 1250 | — | — | — | — | — | — |

PRELIMINARY BLOATING TEST:

| Temperature (°C) | Percent absorption | Bulk density (g/cm³) | (lb/ft³) | Remarks |
|---|---|---|---|---|
| 1000 | 8.5 | 2.07 | 129.3 | Slight expansion. |
| 1050 | 10.1 | 1.82 | 113.8 | do. |
| 1100 | 14.3 | 1.17 | 72.8 | Good pore structure. |
| 1150 | 15.6 | .88 | 54.8 | Good pore structure (sticky). |

Potential use: Structural clay products (e.g., building brick, floor brick) at 1100 to 1150°C. Marginal for lightweight aggregate at 1100 to 1150°C.

Remarks: Short firing range. Might have to be pelletized because of shaly nature.

## Sample 214–D–3

Unit: Mahantango Formation
Quadrangle: Bushkill
Latitude/longitude: 41°05′11″N/75°01′24″W
Sampled interval: 22 feet

RAW PROPERTIES:
Water of plasticity (percent): 15
Color: Gray
pH: 7.6
HCl effervescence: None

Working properties: Short
Drying shrinkage (percent): 2.5
Dry strength: Poor
Preliminary bloating test: Positive

SLOW-FIRING TEST:

| Temperature (°C) | Color[2] | Hardness (Mohs scale) | Percent linear shrinkage | Percent absorption | Percent apparent porosity | Bulk density (g/cm³) |
|---|---|---|---|---|---|---|
| 1000 | 5YR7/6 | 3 | 2.5 | 13.5 | 26.3 | 1.94 |
| 1050 | 2.5YR6/8 | 3 | 5.0 | 10.8 | 22.1 | 2.05 |
| 1100 | 2.5YR4/6 | 3 | 5.0 | 6.8 | 14.9 | 2.09 |
| 1150 | 10R4/4 | 4 | 5.0 | 3.5 | 7.4 | 2.19 |
| 1200 | 10R4/2 | 5 | 5.0 | 1.7 | 3.8 | 2.23 |
| 1250 | — | — | Expanded | — | — | — |

PRELIMINARY BLOATING TEST:

| Temperature (°C) | Percent absorption | Bulk density (g/cm³) | Bulk density (lb/ft³) | Remarks |
|---|---|---|---|---|
| 1000 | 8.5 | 2.05 | 127.8 | Slight expansion. |
| 1050 | 10.6 | 1.83 | 114.4 | do. |
| 1100 | 3.6 | 1.33 | 82.7 | Good pore structure. |
| 1150 | .7 | .90 | 56.4 | do. |

Potential use: Structural clay products (e.g., building brick) at 1150 to 1200 °C. Lightweight aggregate at 1150 °C.

Remarks: Might have to be pelletized because of shaly nature.

## Sample 223–A–1

Unit: Catskill Formation, Lackawaxen Member
Quadrangle: Shohola
Latitude/longitude: 41°25′19″N/74°54′17″W
Sampled interval: 15 feet

RAW PROPERTIES:
Water of plasticity (percent): 17.4
Color: Tan
pH: 7.3
HCl effervescence: None

Working properties: Short
Drying shrinkage (percent): 2.5
Dry strength: Fair
Preliminary bloating test: Negative

Sample 223-A-1 *(Continued)*

SLOW-FIRING TEST:

| Temperature (°C) | Color[2] | Hard-ness (Mohs scale) | Percent linear shrinkage | Percent absorption | Percent apparent porosity | Bulk density (g/cm³) |
|---|---|---|---|---|---|---|
| 1000 | 5YR7/6 | 2 | 2.5 | 17.1 | 31.6 | 1.84 |
| 1050 | 2.5YR6/8 | 3 | 5.0 | 14.1 | 27.4 | 1.94 |
| 1100 | 2.5YR5/6 | 3 | 5.0 | 8.4 | 18.0 | 2.14 |
| 1150 | 10R4/4 | 4 | 7.5 | 3.5 | 8.2 | 2.31 |
| 1200 | 10R4/2 | 5 | 7.5 | 1.0 | 2.5 | 2.40 |
| 1250 | — | — | Melted | — | — | — |

Potential use: Structural clay products (e.g., building brick, floor brick) at 1150 to 1200 °C.

Sample 223-D-1

Unit: Mahantango Formation
Quadrangle: Milford
Latitude/longitude: 41°16′47″N/74°50′23″W
Sampled interval: 26 feet (lower part of exposure)

RAW PROPERTIES:

Water of plasticity (percent): 15.5
Color: Gray
pH: 7.5
HCl effervescence: Slight

Working properties: Short
Drying shrinkage (percent): 0.0
Dry strength: Fair
Preliminary bloating test: Positive

SLOW-FIRING TEST:

| Temperature (°C) | Color[2] | Hard-ness (Mohs scale) | Percent linear shrinkage | Percent absorption | Percent apparent porosity | Bulk density (g/cm³) |
|---|---|---|---|---|---|---|
| 1000 | 5YR7/6 | 2 | 2.5 | 14.8 | 27.9 | 1.89 |
| 1050 | 2.5YR6/8 | 3 | 5.0 | 12.0 | 24.0 | 1.99 |
| 1100 | 2.5YR5/6 | 4 | 5.0 | 7.5 | 16.1 | 2.13 |
| 1150 | 10R4/4 | 4 | 5.0 | 4.7 | 10.4 | 2.22 |
| 1200 | 10R4/2 | 5 | 5.0 | 2.4 | 5.3 | 2.23 |
| 1250 | — | — | Expanded | — | — | — |

PRELIMINARY BLOATING TEST:

| Temperature (°C) | Percent absorption | Bulk density (g/cm³) | Bulk density (lb/ft³) | Remarks |
|---|---|---|---|---|
| 1050 | 9.4 | 1.93 | 120.1 | Slight expansion. |
| 1100 | 9.3 | 1.81 | 113.1 | do. |
| 1150 | 12.4 | .99 | 61.6 | Good pore structure. |
| 1200 | 12.8 | .78 | 48.9 | Some large pores (sticky). |

Potential use: Structural clay products (e.g., building brick, floor brick) at 1100 to 1200 °C. Lightweight aggregate at 1150 to 1200 °C.

Remarks: Might have to be pelletized because of shaly nature. Presence of carbonates could cause problems.

## Sample 223-D-2

Unit: Mahantango Formation
Quadrangle: Milford
Latitude/longitude: 41°16′47″N/74°50′23″W
Sampled interval: 25 feet (upper part of exposure)

RAW PROPERTIES:
Water of plasticity (percent): 14.6
Color: Gray
pH: 7.8
HCl effervescence: Slight

Working properties: Short
Drying shrinkage (percent): 2.5
Dry strength: Fair
Preliminary bloating test: Positive

SLOW-FIRING TEST:

| Temperature (°C) | Color[2] | Hardness (Mohs scale) | Percent linear shrinkage | Percent absorption | Percent apparent porosity | Bulk density (g/cm$^3$) |
|---|---|---|---|---|---|---|
| 1000 | 2.5YR6/8 | 3 | 2.5 | 14.2 | 27.5 | 1.93 |
| 1050 | 2.5YR5/8 | 3 | 2.5 | 11.3 | 23.0 | 2.04 |
| 1100 | 2.5YR4/6 | 4 | 5.0 | 7.5 | 16.3 | 2.17 |
| 1150 | 10R4/4 | 4 | 5.0 | 4.1 | 9.2 | 2.23 |
| 1200 | 10R4/2 | 5 | 5.0 | 2.3 | 5.1 | 2.26 |
| 1250 | — | — | Expanded | — | — | — |

PRELIMINARY BLOATING TEST:

| Temperature (°C) | Percent absorption | Bulk density (g/cm$^3$) | Bulk density (lb/ft$^3$) | Remarks |
|---|---|---|---|---|
| 1050 | 8.3 | 1.95 | 121.6 | Slight expansion. |
| 1100 | 11.8 | 1.38 | 86.4 | Good pore structure. |
| 1150 | 14.1 | .97 | 60.3 | do. |
| 2000 | 16.3 | .77 | 47.9 | Some large pores. |

Potential use: Structural clay products (e.g., building brick, floor brick) at 1100 to 1200°C. Lightweight aggregate at 1150 to 1200°C.

Remarks: Might have to be pelletized because of shaly nature. Presence of carbonates could cause problems.

## Sample 223-D-3

Unit: Mahantango Formation
Quadrangle: Milford
Latitude/longitude: 41°20′44″N/74°45′32″W
Sampled interval: 12 feet

RAW PROPERTIES:
Water of plasticity (percent): 12
Color: Gray
pH: 8.1
HCl effervescence: Slight

Working properties: Short
Drying shrinkage (percent): 2.5
Dry strength: Poor
Preliminary bloating test: Positive

Sample 223-D-3 *(Continued)*

SLOW-FIRING TEST:

| Temperature (°C) | Color[2] | Hardness (Mohs scale) | Percent linear shrinkage | Percent absorption | Percent apparent porosity | Bulk density ($g/cm^3$) |
|---|---|---|---|---|---|---|
| 1000 | 2.5YR6/6 | 3 | 2.5 | 13.0 | 25.8 | 1.99 |
| 1050 | 2.5YR5/6 | 3 | 2.5 | 10.2 | 21.2 | 2.08 |
| 1100 | 2.5YR4/6 | 4 | 5.0 | 6.1 | 13.4 | 2.18 |
| 1150 | 10R3/4 | 5 | 5.0 | 4.0 | 8.6 | 2.20 |
| 1200 | — | — | Expanded | — | — | — |
| 1250 | — | — | — | — | — | — |

PRELIMINARY BLOATING TEST:

| Temperature (°C) | Percent absorption | Bulk density ($g/cm^3$) | Bulk density ($lb/ft^3$) | Remarks |
|---|---|---|---|---|
| 1050 | 8.5 | 1.95 | 121.9 | Slight expansion. |
| 1100 | 11.3 | 1.50 | 93.6 | Partial expansion. |
| 1150 | 15.4 | .74 | 45.9 | Good pore structure. |
| 1200 | 17.2 | .63 | 39.1 | Some large pores. |

Potential use: Structural clay products (e.g., building brick, floor brick) at 1100 to 1150°C. Lightweight aggregate at 1150 to 1200°C.

Remarks: Presence of carbonates could cause problems.

Sample 223-D-4

Unit: Mahantango Formation
Quadrangle: Milford
Latitude/longitude: 41°20′42″N/74°45′53″W
Sampled interval: 20 feet

RAW PROPERTIES:
Water of plasticity (percent): 12.5
Color: Gray
pH: 8.2
HCl effervescence: Slight
Working properties: Short
Drying shrinkage (percent): 2.5
Dry strength: Poor
Preliminary bloating test: Positive

SLOW-FIRING TEST:

| Temperature (°C) | Color[2] | Hardness (Mohs scale) | Percent linear shrinkage | Percent absorption | Percent apparent porosity | Bulk density ($g/cm^3$) |
|---|---|---|---|---|---|---|
| 1000 | 2.5YR6/8 | 3 | 2.5 | 12.4 | 24.8 | 2.01 |
| 1050 | 2.5YR6/6 | 3 | 5.0 | 9.8 | 20.4 | 2.09 |
| 1100 | 2.5YR4/6 | 4 | 5.0 | 4.6 | 10.4 | 2.22 |
| 1150 | 10R4/4 | 5 | 5.0 | 2.6 | 5.8 | 2.27 |
| 1200 | — | — | Expanded | — | — | — |
| 1250 | — | — | — | — | — | — |

## Sample 223-D-4 *(Continued)*

PRELIMINARY BLOATING TEST:

| Temperature (°C) | Percent absorption | Bulk density (g/cm³) | Bulk density (lb/ft³) | Remarks |
|---|---|---|---|---|
| 1050 | 10.4 | 1.83 | 114.0 | Slight expansion. |
| 1100 | 4.7 | 1.26 | 78.9 | Good pore structure. |
| 1150 | 12.9 | .75 | 46.6 | do. |
| 1200 | 12.5 | .59 | 36.9 | Some large pores. |

Potential use: Structural clay products (e.g., building brick, floor brick) at 1100 to 1150 °C. Lightweight aggregate at 1150 to 1200 °C.

Remarks: Presence of carbonates could cause problems.

## Sample 223-D-5

Unit: Catskill Formation, Towamensing Member
Quadrangle: Milford
Latitude/longitude: 41°20′15″N/74°52′01″W
Sampled interval: 22 feet

RAW PROPERTIES:

Water of plasticity (percent): 16.1
Color: Gray
pH: 8.2
HCl effervescence: Slight

Working properties: Short
Drying shrinkage (percent): 2.5
Dry strength: Fair
Preliminary bloating test: Positive

SLOW-FIRING TEST:

| Temperature (°C) | Color[2] | Hardness (Mohs scale) | Percent linear shrinkage | Percent absorption | Percent apparent porosity | Bulk density (g/cm³) |
|---|---|---|---|---|---|---|
| 1000 | 2.5YR6/8 | 3 | 2.5 | 14.3 | 27.8 | 1.94 |
| 1050 | 2.5YR6/6 | 3 | 5.0 | 11.0 | 22.6 | 2.06 |
| 1100 | 2.5YR5/6 | 4 | 5.0 | 5.4 | 12.3 | 2.28 |
| 1150 | 2.5YR4/4 | 5 | 7.5 | 1.7 | 4.2 | 2.40 |
| 1200 | — | — | Expanded | — | — | — |
| 1250 | — | — | — | — | — | — |

PRELIMINARY BLOATING TEST:

| Temperature (°C) | Percent absorption | Bulk density (g/cm³) | Bulk density (lb/ft³) | Remarks |
|---|---|---|---|---|
| 1050 | 4.3 | 2.03 | 126.7 | Slight expansion. |
| 1100 | 7.8 | 1.36 | 84.9 | Good pore structure. |
| 1150 | 7.0 | 1.12 | 69.9 | do. |
| 1200 | 11.5 | .75 | 46.8 | Some large pores (sticky). |

Potential use: Structural clay products (e.g., building brick, floor brick) at 1100 to 1150 °C. Lightweight aggregate at 1150 to 1200 °C.

Remarks: Presence of carbonates could cause problems.

## Sample 224–A–1

Unit: Mahantango Formation
Quadrangle: Lake Maskenoza
Latitude/longitude: 41°10′55″N/74°53′33″W
Sampled interval: 32 feet (lower part of exposure)

RAW PROPERTIES:
Water of plasticity (percent): 12.5
Color: Gray
pH: 7.9
HCl effervescence: Slight
Working properties: Short
Drying shrinkage (percent): 0.0
Dry strength: Poor
Preliminary bloating test: Negative

SLOW-FIRING TEST:

| Temperature (°C) | Color [2] | Hardness (Mohs scale) | Percent linear shrinkage | Percent absorption | Percent apparent porosity | Bulk density (g/cm$^3$) |
|---|---|---|---|---|---|---|
| 1000 | 2.5YR6/8 | 3 | 2.5 | 13.1 | 26.0 | 1.98 |
| 1050 | 2.5YR5/8 | 3 | 2.5 | 9.6 | 20.1 | 2.10 |
| 1100 | 2.5YR5/6 | 4 | 5.0 | 5.4 | 12.2 | 2.26 |
| 1150 | 10R4/4 | 4 | 5.0 | 3.0 | 7.0 | 2.30 |
| 1200 | 10R4/2 | 5 | 5.0 | 1.0 | 2.3 | 2.33 |
| 1250 | — | — | Melted | — | — | — |

Potential use: Structural clay products (e.g., building brick, floor brick) at 1100 to 1200°C.

Remarks: Presence of carbonates could cause problems.

## Sample 224–A–2

Unit: Mahantango Formation
Quadrangle: Lake Maskenoza
Latitude/longitude: 41°10′55″N/74°53′33″W
Sampled interval: 32 feet (upper part of exposure)

RAW PROPERTIES:
Water of plasticity (percent): 12.2
Color: Gray
pH: 8.8
HCl effervescence: Slight
Working properties: Short
Drying shrinkage (percent): 0.0
Dry strength: Poor
Preliminary bloating test: Positive

SLOW-FIRING TEST:

| Temperature (°C) | Color [2] | Hardness (Mohs scale) | Percent linear shrinkage | Percent absorption | Percent apparent porosity | Bulk density (g/cm$^3$) |
|---|---|---|---|---|---|---|
| 1000 | 2.5YR6/8 | 3 | 2.5 | 13.2 | 26.0 | 1.97 |
| 1050 | 2.5YR5/8 | 3 | 5.0 | 9.2 | 19.6 | 2.13 |
| 1100 | 2.5YR4/6 | 3 | 5.0 | 5.8 | 13.1 | 2.25 |
| 1150 | 10R4/4 | 4 | 5.0 | 3.2 | 7.6 | 2.33 |
| 1200 | — | — | Expanded | — | — | — |
| 1250 | — | — | — | — | — | — |

Sample 224-A-2 *(Continued)*

PRELIMINARY BLOATING TEST:

| Temperature (°C) | Percent absorption | Bulk density ($g/cm^3$) | ($lb/ft^3$) | Remarks |
|---|---|---|---|---|
| 1050 | 10.0 | 1.85 | 115.5 | Slight expansion. |
| 1100 | 7.0 | 1.51 | 94.5 | Partial expansion. |
| 1150 | 12.1 | 1.05 | 65.3 | Good pore structure. |
| 1200 | 13.6 | .79 | 49.4 | Good pore structure (sticky). |

Potential use: Marginal for structural clay products (e.g, building brick) at 1150 °C. Lightweight aggregate at 1150 to 1200 °C.

Remarks: Short firing range. Presence of carbonates could cause problems.